Martingale Inequalities
Seminar Notes on Recent Progress

MATHEMATICS LECTURE NOTE SERIES

Published

J. Frank Adams	Lectures on Lie Groups, 1969
E. Artin and J. Tate	Class Field Theory, 1967
Michael F. Atiyah	K-Theory, 1967
Jacob Barshay	Topics in Ring Theory, 1969
Hyman Bass	Algebraic K-Theory, 1968
Melvyn Berger and Marion Berger	Perspectives in Nonlinearity: An Introduction to Nonlinear Analysis, 1968
Armand Borel	Linear Algebraic Groups, 1969
Raoul Bott	Lectures on K(X), 1969
Andrew Browder	Introduction to Function Algebras, 1969
John L. Challifour	Generalized Functions and Fourier Analysis, 1972
Gustave Choquet	Lectures on Analysis
	Volume I. Integration and Topological Vector Spaces, 1969
	Volume II. Representation Theory, 1969
	Volume III. Infinite Dimensional Measures and Problem Solutions, 1969
Paul J. Cohen	Set Theory and the Continuum Hypothesis, 1966
Eldon Dyer	Cohomology Theories, 1969
Robert Ellis	Lectures on Topological Dynamics, 1969
Walter Feit	Characters of Finite Groups, 1967
John Fogarty	Invariant Theory, 1969
William Fulton	Algebraic Curves: An Introduction to Algebraic Geometry, 1969
Adriano M. Garsia	Martingale Inequalities: Seminar Notes on Recent Progress, 1973
Marvin J. Greenberg	Lectures on Algebraic Topology, 1967
Marvin J. Greenberg	Lectures on Forms in Many Variables, 1969
Robin Hartshorne	Foundations of Projective Geometry, 1967
Robert Hermann	Fourier Analysis on Groups and Partial Wave Analysis, 1969
Robert Hermann	Lectures in Mathematical Physics, Volume I, 1970 Volume II, 1972
Robert Hermann	Lie Algebras and Quantum Mechanics, 1970

J.F.P. Hudson	Piecewise Linear Topology, 1969
Irving Kaplansky	Rings of Operators, 1968
Kenneth M. Kapp and Hans Schneider	Completely O-Simple Semigroups: An Abstract Treatment of the Lattice of Congruences, 1969
Joseph B. Keller and Stuart Antman (Editors)	Bifurcation Theory and Nonlinear Eigenvalue Problems, 1969
Irwin Kra	Automorphic Forms and Kleinian Groups, 1972
T. Y. Lam	The Algebraic Theory of Quadratic Forms, 1973
Serge Lang	Algebraic Functions, 1965
Serge Lang	Rapport sur la cohomologie des groupes, 1966
Ottmar Loos	Symmetric Spaces Volume I. General Theory, 1969 Volume II. Compact Spaces and Classifications, 1969
I.G. Macdonald	Algebraic Geometry: Introduction to Schemes, 1968
George W. Mackey	Induced Representations of Groups and Quantum Mechanics, 1968
Hideyuki Matsumura	Commutative Algebra, 1969
Richard K. Miller	Nonlinear Volterra Integral Equations, 1971
Andrew Ogg	Modular Forms and Dirichlet Series, 1969
Richard S. Palais	Foundations of Global Non-Linear Analysis, 1968
William Parry	Entropy and Generators in Ergodic Theory, 1969
Donald Passman	Permutation Groups, 1968
Walter Rudin	Function Theory in Polydiscs, 1969
David Russell	Optimization Theory, 1970
Gerald E. Sacks	Saturated Model Theory, 1972
Jean-Pierre Serre	Abelian l-Adic Representations and Elliptic Curves, 1968
Jean-Pierre Serre	Algèbres de Lie semi-simples complexes, 1966
Jean-Pierre Serre	Lie Algebras and Lie Groups, 1965
Shlomo Sternberg	Celestial Mechanics Part I, 1969
Shlomo Sternberg	Celestial Mechanics Part II, 1969
Moss E. Sweedler	Hopf Algebras, 1969

Martingale Inequalities
Seminar Notes on Recent Progress

Adriano M. Garsia
University of California, San Diego

1973
W. A. BENJAMIN, INC.
ADVANCED BOOK PROGRAM
Reading, Massachusetts

London · Amsterdam · Don Mills, Ontario · Sydney · Tokyo

Garsia, Adriano   M   1928–
Martingale Inequalities: Seminar Notes on Recent Progress

(Mathematics lecture note series)
Includes bibliography: p.
I. Title
                                                    73-7952
ISBN 0-805-33102-6
ISBN 0-805-33103-4 (pbk)

Reproduced by W. A. Benjamin, Inc., Advanced Book Program, Reading, Massachusetts, from camera-ready copy prepared by the author.

American Mathematical Society (MOS) Subject Classification Scheme (1970):
60G45, 25A86

Copyright © 1973 by W. A. Benjamin, Inc.
Philippines copyright 1973 by W. A. Benjamin, Inc.
Published simultaneously in Canada.

All rights reserved. No part of this publication may be reproduced, stored in a retrieval system, or transmitted, in any form or by any means, electronic, mechanical, photocopying, recording, or otherwise, without the prior written permission of the publisher, W. A. Benjamin, Inc., Advanced Book Program.

Manufactured in the United States of America

ISBN 0-805-33102-6 (hardbound)
ISBN 0-805-33103-4 (paperback)
ABCDEFGHIJ-HA-7876543

TABLE OF CONTENTS

PREFACE 1

I DUALITY OF $\mathcal{H}_p$ AND $\mathcal{K}_q$ SPACES

1. The $\mathcal{H}_p$ and BMO norms 3
2. The $\mathcal{K}_q$ norm 4
3. Fefferman's inequality 6
4. A representation for the functionals on $\mathcal{H}_p$ 12
5. A remarkable representation for BMO functions 22

II INEQUALITIES BETWEEN THE SQUARE FUNCTION AND THE MAXIMAL FUNCTION

1. The equivalence of $\mathcal{H}_q$ and $\mathcal{K}_q$ norms 27
2. Proof of the inequalities for $1 \leq p \leq 2$ 34
3. Proofs of the auxiliary lemmas 41
4. A further characterization of BMO functions 48
5. Square function inequalities for supermartingales and Burkholder's weak-L_1 inequality 58

III BMO SEQUENCES AND POTENTIALS

1. Exponential bounds for monotone sequences 64
2. Proof of the John-Nirenberg theorem 75
3. $\mathcal{H}_1$, the L log L class and an inequality of Doob 81

4. Potentials and a "dual" to Doob's inequality 97
5. On the possibility of extending the definitions of $\mathcal{K}_q$ spaces 114

IV MARTINGALE TRANSFORM TECHNIQUES

1. The space $\mathcal{H}_1^-$ and its relation to $\mathcal{P}$ and $\mathcal{H}_1$ 124
2. The spaces $\mathcal{P}_p$, $\mathcal{K}_p^+$ and a further generalization of Fefferman's inequality 130
3. L_p estimates between f^* and the conditional square function . . . 142
4. Relations amongst $\mathcal{P}_p$ spaces and further remarks. 148

NOTATION AND BASIC DEFINITIONS 163
INDEX TO THE INEQUALITIES 167
NOTEWORTHY RESULTS AND REMARKS 173
BIBLIOGRAPHY 181

Preface

The material contained in these notes was put together as a result of some lectures delivered at the analysis seminar in La Jolla in the Fall of 1971. The object of these lectures was to present the recent work of C. Fefferman on functions of bounded mean oscillation and its relation to some work of Burkholder, Davis and Gundy on martingales.

In preparing these lectures an effort was made to establish the various inequalities by the simplest and most direct path trying whenever possible to obtain reasonable expressions for the constants involved.

Some of the methods of proof that thereby were developed turn out to be of interest in themselves, and it seems worthwhile that they be available to a wider audience.

The present notes cover only the "martingale" portion of the seminar lectures. For the "harmonic function" portion the reader is referred to the preliminary notes [28] or the original papers [12], [13] and [9].

The contents are divided into four chapters. Each chapter starts with a basic result as a goal and then further material is presented which stems from the methods used in the proof of that result.

The goal of the first Chapter is Fefferman's theorem. This leads to a new "L_q" norm for martingales which for $q < \infty$ is equivalent to the ordinary L_q norm and for $q = \infty$ reduces to the BMO norm. An extension of Fefferman's inequality is thereby obtained which is crucial in deriving the results of the second chapter.

The goal of the second chapter are the L_p estimates involving the "maximal functions" and the "square function". Our proof here, which is based on our generalization of Fefferman's inequality works equally well for the case $p > 1$ (Gundy and Burkholder [8]) and for $p = 1$

(*) This research was carried out under support of Grant AF-AFOSR 2088.

(Burgess Davis [10]). In the process some interesting representations for BMO and H_p functions are also obtained.

The third chapter is dedicated to the proof of the John-Nirenberg theorem and ramifications thereof. This leads to certain basic estimates concerning potentials.

Our aim in the fourth chapter is to introduce certain "martingale transform" methods. These methods can be viewed as an extension of the classical "stopping time" arguments.

Roughly speaking, whenever a proof of a certain estimate on martingales is obtained by a combination of stopping time argument (to get a distribution function inequality) followed by an integration, a proof of a usually sharper estimate can be derived at once by means of a martingale transform argument.

These methods are used in the fourth chapter mostly to derive relations amongst the various spaces we deal with here. In particular, the L_p estimates between the maximal function and the conditioned square function are shown there to be direct consequences of the Parseval identity for martingales.

Throughout these notes we have added remarks and observations in the hope that they might be helpful for further research. These are too numerous to be listed in detail. However, we have included an index at the end of the notes, where our basic definitions and theorems are listed, to help the reader who wishes a more accurate description of the contents.

Finally, we would like to acknowledge here our gratitude to R. Getoor and M. Sharpe for being such a patient, critical and informing audience in our endless mathematical conversations on this subject. This provided us with an invaluable stimulus towards the completion of our task.

Adriano M. Garsia

I. DUALITY OF $\mathcal{H}_p$ AND $\mathcal{K}_q$ SPACES.

I.1 The $\mathcal{H}_p$ and BMO norms.

Throughout these notes we shall work with a fixed probability space $(\Omega, \mathcal{F}, P)$ and a sequence of σ-fields

$$\mathcal{F}_0 \subset \mathcal{F}_1 \subset \ldots \subset \mathcal{F}_n \subset \ldots \subset \mathcal{F}$$

such that $\bigvee_{n=1}^{\infty} \mathcal{F}_n = \mathcal{F}$.

For a random variable $f \in L_1(\Omega, \mathcal{F}, P)$ we shall set

$$f_n = E(f|\mathcal{F}_n), \quad \Delta f_n = f_n - f_{n-1},$$

$$f_n^* = \max_{0 \le \nu \le n} |f_\nu|, \quad f^* = \sup_n f_n^*,$$

$$S_n(f) = \sqrt{\sum_{\nu=1}^{n} [\Delta f_\nu]^2}, \quad S(f) = \sup_n S_n(f).$$

We shall also introduce the spaces

I.1.1 $\qquad \mathcal{H}_p = \{f : E([S(f)]^p) < \infty\} \quad (p \ge 1).$

with norm

$$\|f\|_{\mathcal{H}_p} = [E([S(f)]^p)]^{1/p}$$

Furthermore we let

I.1.2 $\quad$ BMO $= \{f: \sup_{n \geq 1} \| E(|f - f_{n-1}|^2 | \mathcal{F}_n) \|_\infty < \infty \}$.

with norm

$$\|f\|_{BMO} = \sup_{n \geq 1} \| \sqrt{E(|f - f_{n-1}|^2 | \mathcal{F}_n)} \|_\infty. \quad (*)$$

The reason behind this terminology lies in the fact that, with an apropriate choice of $(\Omega, \mathcal{F}, P)$ and $\{\mathcal{F}_n\}$, the spaces in I.1.1 can be identified with the classical $\mathcal{H}_p$ spaces of function theory while at the same time the space defined in I.1.2 can be identified with the class of functions of bounded mean oscillation introduced by John and Nirenberg.

I.2 The $\mathcal{K}_q$ norm.

Parallel to the $\mathcal{H}_p$ spaces it is convenient to consider what we shall call "$\mathcal{K}_q$" spaces ($q \geq 2$) with norm $\|f\|_{\mathcal{K}_q}$ defined as follows.

Given a martingale $f_n = E(f|\mathcal{F}_n)$ ($f \in L_2$) we shall denote by Γ_f the class of functions γ such that

---

(*) Strictly speaking, these functionals are norms in the usual sense only when restricted to f's such that $f_0 = E(f|\mathcal{F}_0) = 0$.

I.2.1 $\quad E(|f - f_{n-1}|^2 | \mathcal{F}_n) \leq E(\gamma^2 | \mathcal{F}_n) \quad \forall \, n .$

We shall then set

$$\|f\|_{\mathcal{K}_q} = \inf_{\gamma \in \Gamma_f} [E(\gamma^q)]^{1/q}$$

and let

I.2.2 $\quad \mathcal{K}_q = \{f: \|f\|_{\mathcal{K}_q} < \infty\} .$

Remark I.1.1 We want to observe once and for all that for $\varphi \in L_2$

I.2.3 $\quad E(|\varphi - \varphi_{n-1}|^2 | \mathcal{F}_n) = \sum_{\nu=n}^{\infty} E(|\Delta \varphi_\nu|^2 | \mathcal{F}_n)$

Indeed, if $\mu - 1 \geq \nu$

$$E(\Delta \varphi_\mu | \mathcal{F}_\nu) = E(\varphi_\mu | \mathcal{F}_\nu) - E(\varphi_{\mu-1} | \mathcal{F}_\nu) = \varphi_\nu - \varphi_\nu = 0 .$$

Thus, if $f \in L_p$ and $\varphi \in L_q$ ($q = \frac{p}{p-1}$) we have for $\mu > \nu \geq n$

$$E(\Delta f_\nu \Delta \varphi_\mu | \mathcal{F}_n) = E(\Delta f_\nu E(\Delta \varphi_\mu | \mathcal{F}_\nu) | \mathcal{F}_n) = 0 \ .$$

This gives immediately

I.2.4 $\quad E\left((f_N - f_{n-1})(\varphi_N - \varphi_{n-1})|\mathcal{F}_n\right) = \sum_{\nu=n}^{N} E(\Delta f_\nu \Delta \varphi_\nu | \mathcal{F}_n) ,$

and I.2.3 follows for $f = \varphi$.

In particular we see that if $\varphi \in$ BMO then

$$[\Delta \varphi_n]^2 \leq E(|\varphi - \varphi_{n-1}|^2 | \mathcal{F}_n) \leq \|\varphi\|_{BMO}^2$$

Thus for a BMO function φ all φ_n's are bounded.

Going back to I.2.2 we see that

$$BMO = \mathcal{K}_\infty$$

and indeed $\|f\|_{BMO} = \|f\|_{\mathcal{K}_\infty}$.

I.3 Fefferman's inequality.

Quite recently C. Fefferman discovered the remarkable fact that BMO is none other than the "dual" of $\mathcal{H}_1$ in the sense of functional analysis.

This result, of course, consists of two parts namely:

a) An inequality, (here and after referred to as "Fefferman's inequality"), which can be written "formally" as

I.3.1 $$|E(f\varphi)| \leq c \, \|f\|_{H_1} \, \|\varphi\|_{BMO}.$$

b) a theorem to the effect that every functional $L(f)$ such that

$$|L(f)| \leq B\|f\|_{H_1}$$

must necessarily be of the "form"

I.3.2 $$L(f) = E(f\varphi)$$

with $\|\varphi\|_{BMO} \leq c\,B$.

We used the word "formally" since $E(f\varphi)$ need not make sense as a "Lebesgue" integral when we only know that $f \in H_1$ and $\varphi \in BMO$. However, one of the by-products of Fefferman's theorem is that we can define $E(f\varphi)$ by setting

$$E(f\varphi) = \lim_{n \to \infty} E(f_n \varphi_n).$$

It is in this manner that $E(f\varphi)$ is to be interpreted in I.3.1 and I.3.2.

Our goal in this section will be to prove the corresponding results concerning $\mathcal{H}_p$ ($1 \leq p \leq 2$) and $\mathcal{K}_q$ ($q = \frac{p}{p-1}$), thereby obtaining Fefferman's theorem as a particular case.

Our first inequality can be stated as follows:

<u>Theorem 1.3.1</u> <u>Let</u> $f \in \mathcal{H}_p$ $1 \leq p \leq 2$ <u>and</u> $\varphi \in \mathcal{K}_q$ ($q = \frac{p}{p-1}$) <u>then</u>, <u>assuming</u> $f_0 = \varphi_0 = 0$, <u>we have</u>

I.3.3
$$|E(f_n \varphi_n)| \leq \sqrt{\frac{2}{p}}\ \|f\|_{\mathcal{H}_p}\ \|\varphi\|_{\mathcal{K}_q}.$$

<u>Proof.</u> Before we can proceed with the argument, some remarks are in order. First of all, since

$$|f_n| = \left|\sum_{\nu=1}^{n} \Delta f_\nu\right| \leq \sqrt{n}\ \sqrt{\sum_{\nu=1}^{n} [\Delta f_\nu]^2} = \sqrt{n}\ S_n(f),$$

we see that $S(f) \in L_p$ implies $f_n \in L_p$ $\forall\ n$.

Secondly if $\varphi \in \mathcal{K}_q$, in view of I.2.3 for some $\gamma \in L_q$ we have

I.3.4
$$|\Delta\varphi_n|^2 \leq E(|\varphi - \varphi_{n-1}|^2\ |\mathcal{F}_n) \leq E(\gamma^2|\mathcal{F}_n).$$

Thus, each φ_n is also in L_q and we can at least conclude that the left hand side of I.3.3 makes sense for all n and is given by

I.3.5 $$E(f_n \varphi_n) = \sum_{\nu=1}^{n} E(\Delta f_\nu \, \Delta \varphi_\nu) \, .$$

To obtain I.3.3 we start with a trick suggested by a proof of Fefferman's inequality due to Carl Herz. This consists in writing I.3.5 in the form

I.3.6 $$E(f_n \varphi_n) = \sum_{\nu=1}^{n} E\left(\frac{\Delta f_\nu}{s_\nu^{1-p/2}} \, s_\nu^{1-p/2} \Delta \varphi_\nu\right)$$

and then applying Schwarz's inequality to obtain

I.3.7 $$|E(f_n \varphi_n)| \le \sqrt{A} \, \sqrt{B}$$

where

$$A = \sum_{\nu=1}^{n} E\left(\frac{[\Delta f_\nu]^2}{s_\nu^{2-p}}\right),$$

$$B = \sum_{\nu=1}^{n} E\left(s_\nu^{2-p} [\Delta \varphi_\nu]^2\right).$$

This given, the proof of I.3.3 can proceed as follows. First of all, we note that

$$\frac{S_\nu^2 - S_{\nu-1}^2}{S_\nu^{2-p}} \leq \frac{2}{p} (S_\nu^p - S_{\nu-1}^p). \quad (*)$$

Thus we immediately deduce

I.3.8 $$A = \sum_{\nu=1}^\infty E\left(\frac{S_\nu^2 - S_{\nu-1}^2}{S_\nu^{2-p}}\right) \leq \frac{2}{p} E(S^p).$$

To estimate B, for convenience, we set

$$\theta_\nu = S_\nu^{2-p} - S_{\nu-1}^{2-p}$$

and note that, since S_ν increases and $p \leq 2$, each θ_ν is non negative.

We can then write

(*) For $\alpha \leq 1 \leq \rho$ we have $(\rho^\alpha - 1) \geq \alpha(\rho - 1) \rho^{\alpha-1}$, then set $\rho = S_\nu^2/S_{\nu-1}^2$ and $\alpha = p/2$.

$$B = \sum_{\nu=1}^{n} \sum_{\mu=1}^{\nu} E(\theta_\mu [\Delta\varphi_\nu]^2) =$$

$$= \sum_{\mu=1}^{n} \sum_{\nu=\mu}^{n} E(\theta_\mu [\Delta\varphi_\nu]^2) =$$

$$= \sum_{\mu=1}^{n} E\left(\theta_\mu E\left(\sum_{\nu=\mu}^{n} [\Delta\varphi_\nu]^2 \,\big|\, \mathcal{F}_\mu\right)\right) =$$

$$\leq \sum_{\mu=1}^{n} E\left(\theta_\mu E(|\varphi - \varphi_{\mu-1}|^2 \,|\, \mathcal{F}_\mu)\right).$$

Now, if $p = 1$ then we simply estimate $E(|\varphi - \varphi_{\mu-1}|^2 \,|\, \mathcal{F}_\mu)$ by $\|\varphi\|_{BMO}^2$ and get

$$B \leq E(S_n(f)) \, \|\varphi\|_{BMO}^2 .$$

In case $p > 1$, we use I.3.4 to derive

$$B \leq \sum_{\mu=1}^{n} E\left(\theta_\mu E(\gamma^2 \,|\, \mathcal{F}_\mu)\right) = E(S_n^{2-p} \gamma^2).$$

So by Hölder's inequality we obtain

$$B \leq \left[E(S_n^p)\right]^{\frac{2}{p}-1} \left[E\left(\gamma^{\frac{p}{p-1}}\right)\right]^{\frac{2(p-1)}{p}} .$$

Combining with I.3.7 and I.3.8 we get

$$|E(f_n \varphi_n)| \leq \sqrt{\frac{2}{p}} \left[E(S^p)\right]^{\frac{1}{2}} \left[E(S^p)\right]^{\frac{1}{p}-\frac{1}{2}} \left[E(\gamma^q)\right]^{\frac{1}{q}}$$

and this of course implies I.3.3.

I.4 A representation for the functionals on $\mathcal{H}_p$.

To obtain his representation result C. Fefferman observed that $\mathcal{H}_1$ can be naturally imbedded into a larger space whose dual is much easier to characterize. We shall use here the same basic observation. However, before we can proceed with our arguments we need to present some auxiliary material.

Let $\mathcal{SH}_p$ denote the Banach space of sequences of random variables

$$\theta = (\theta_1, \theta_2, \ldots, \theta_n, \ldots)$$

with norm

$$\|\theta\|_{\mathcal{SH}_p} = \left[E\left(\left[\sum_{\nu=1}^{\infty} \theta_\nu^2\right]^{p/2}\right)\right]^{1/p}.$$

Lemma I.4.1 *If* $\Lambda(\theta)$ *is a linear functional on* $\mathcal{SH}_p$ *such that*

I.4   13

I.4.1 $\qquad |\Lambda(\theta)| \leq B \, \|\theta\|_{S\mathcal{H}_p} \quad \forall \; \theta \in S\mathcal{H}_p$

then there is a $\sigma \in S\mathcal{H}_q$ with

I.4.2 $\qquad \|\sigma\|_{S\mathcal{H}_q} \leq B$

such that

I.4.3 $\qquad \Lambda(\theta) = \sum_{\nu=1}^{\infty} E(\sigma_\nu \, \gamma_\nu) \; .$

Proof. This result is elementary but for sake of completeness we might as well outline a proof.

Let us assume first that $p > 1$ and set for all $f \in L_p$

$$\Lambda_n(f) = \Lambda(\theta)$$

where

$$\theta_\nu = \begin{cases} f & \text{for} \quad \nu = n \, , \\ 0 & \text{for} \quad \nu \neq n \, . \end{cases}$$

Then I.4.1 gives

$$|\Lambda_n(f)| \leq B \; [E(|f|^p)]^{1/p}.$$

Thus there is a $\sigma_n \subset L_q$ $(q = \frac{q}{q-1})$ such that

$$\Lambda_n(f) = E(\sigma_n f).$$

By linearity we can thus conclude that, at least when $\theta_\nu = 0 \quad \forall \; \nu > n$

$$\Lambda(\theta) = \sum_{\nu=1}^{n} E(\sigma_\nu \theta_\nu).$$

This given, if we choose

$$\theta_\nu = \sigma_\nu \left[\sum_{\mu=1}^{n} \sigma_\mu^2\right]^{(q-2)/2},$$

I.4.1 yields

$$E\left(\left[\sum_{\nu=1}^{n} \sigma_\nu^2\right]^{q/2}\right) \leq B \left[E\left(\left[\sum_{\nu=1}^{n} \sigma_\nu^2\right]^{(q-1)p/2}\right)\right]^{1/p}.$$

but since $(q-1)p = q$ we derive

$$\left[E\left(\left[\sum_{\nu=1}^{n}\sigma_{\nu}^{2}\right]^{q/2}\right)\right]^{1/q} \leq B .$$

Passing to the limit as $n \to \infty$ we obtain I.4.2.

The result for $p = 1$ is obtained by observing that $\mathcal{SH}_p \subset \mathcal{SH}_1$ for $p > 1$. Thus, a $\sigma = (\sigma_1, \sigma_2, \ldots, \sigma_n, \ldots)$ must necessarily exist satisfying I.4.3 $\forall \theta \in \mathcal{SH}_p$ and I.4.2 $\forall q < \infty$. This given, we need only pass to the limit as $q \to \infty$ to obtain

$$\left\|\left[\sum_{\nu=1}^{\infty}\sigma_{\nu}^{2}\right]^{1/2}\right\|_{\infty} \leq B$$

as asserted.

The next auxiliary result is

<u>Lemma I.4.2</u> <u>Let</u> $g_n = E(g|\mathcal{F}_n)$ <u>where</u> $g \in L_q$ $(q > 1)$ <u>and let</u> $g^* = \sup_n |g_n|$. <u>Then</u>

$$\left[E([g^*]^q)\right]^{1/q} \leq \frac{q}{q-1}\left[E(|g|^q)\right]^{1/q}$$

This result is classical, see [11] for a proof. Also, later in these notes a rather simple proof will be given of a much more general theorem. Thus the proof will be omitted here.

Finally, we have the rather interesting

Theorem I.4.1 <u>Let</u> $q \geq 2$ <u>and</u> $\sigma = (\sigma_1, \sigma_2, \ldots, \sigma_n, \ldots)$ <u>be a sequence of random variables such that</u>

$$\text{I.4.4} \qquad \left[E\left(\left[\sum_{\nu=1}^{\infty} \sigma_\nu^2 \right]^{q/2} \right) \right]^{1/q} \leq B$$

<u>then, the martingale</u>

$$\varphi_n = \sum_{\nu=1}^{n} \left[E(\sigma_\nu | \mathcal{F}_\nu) - E(\sigma_\nu | \mathcal{F}_{\nu-1}) \right]$$

<u>converges to a function</u> φ <u>in</u> $\mathcal{K}_q$ <u>such that</u>

$$\text{I.4.5} \qquad \|\varphi\|_{\mathcal{K}_q} \leq \frac{2q}{q-1} B.$$

Proof. Let us recall that to estimate $\|\varphi\|_{\mathcal{K}_q}$ we need only exhibit a function $\gamma \in L_q$ such that

$$\text{I.4.6} \qquad \sum_{\nu=n}^{\infty} E\left([\Delta \varphi_\nu]^2 | \mathcal{F}_n \right) \leq E(\gamma^2 | \mathcal{F}_n) \qquad \forall\ n.$$

We claim we can obtain this with

$$\gamma = 2g^*$$

where

$$g = \left[\sum_{\nu=1}^{\infty} \sigma_\nu^2\right]^{1/2}, \quad g^* = \sup_n E(g|\mathcal{F}_n).$$

Clearly we need only show

I.4.7 $$\sum_{\nu=n}^{N} E([\Delta\varphi_\nu]^2|\mathcal{F}_n) \leq E(\gamma^2|\mathcal{F}_n) \quad \forall \ n \leq N.$$

Thus, to avoid convergence problems let us assume that all σ_ν, with $\nu > N$, vanish identically.

This given, we see that if we set

$$\varphi = \sum_{\nu=1}^{\infty} \left[E(\sigma_\nu|\mathcal{F}_\nu) - E(\sigma_\nu|\mathcal{F}_{\nu-1})\right]$$

then

$$\Delta\varphi_\nu = E(\sigma_\nu|\mathcal{F}_\nu) - E(\sigma_\nu|\mathcal{F}_{\nu-1}).$$

Thus for all $\nu \geq n+1$

$$E([\Delta\varphi_\nu]^2|\mathcal{F}_n) = E\Big(E([\Delta\varphi_\nu]^2|\mathcal{F}_{\nu-1})\Big|\mathcal{F}_n\Big) =$$
$$= E\Big(\big[E(\sigma_\nu|\mathcal{F}_\nu)\big]^2 - \big[E(\sigma_\nu|\mathcal{F}_{\nu-1})\big]^2\Big|\mathcal{F}_n\Big) \leq$$
$$\leq E\Big(E(\sigma_\nu^2|\mathcal{F}_\nu)\Big|\mathcal{F}_n\Big) = E(\sigma_\nu^2|\mathcal{F}_n).$$

On the other hand we have

$$[\Delta\varphi_n]^2 = \big[E(\sigma_n|\mathcal{F}_n)\big]^2 - 2\,E(\sigma_n|\mathcal{F}_n)\,E(\sigma_n|\mathcal{F}_{n-1}) + \big[E(\sigma_n|\mathcal{F}_{n-1})\big]^2 \leq$$
$$\leq E(\sigma_n^2|\mathcal{F}_n) + 2\,g^{*2} + g^{*2}.$$

Therefore, combining these estimates we get

$$\sum_{\nu=n}^{\infty} E([\Delta\varphi_\nu]^2|\mathcal{F}_n) \leq \sum_{\nu=n}^{\infty} E(\sigma_\nu^2|\mathcal{F}_n) + 3\,E(g^{*2}|\mathcal{F}_n).$$
$$\leq 4\,E(g^{*2}|\mathcal{F}_n).$$

This establishes our claim.

Now, to complete the proof of the theorem we need only use Lemma I.4.1 and obtain

$$[E(\gamma^q)]^{1/q} \leq 2 [E([g^*]^q)]^{1/q} \leq \frac{2q}{q-1} [E(g^q)]^{1/q},$$

which, in view of I.4.4, yields I.4.5 as asserted.

A most interesting special case of this result is obtained by letting $q = \infty$. We can state it as follows

Theorem I.4.2 Let $(\sigma_1, \sigma_2, \ldots, \sigma_n, \ldots)$ be a sequence of random variables such that

$$\sum_{\nu=1}^{\infty} \sigma_\nu^2 \leq B^2 \quad \text{a.s.}$$

then the function

$$\varphi = \lim_{n \to \infty} \sum_{\nu=1}^{n} [E(\sigma_\nu | \mathcal{F}_\nu) - E(\sigma_\nu | \mathcal{F}_{\nu-1})]$$

is BMO and indeed

$$\|\varphi\|_{BMO} \leq 2B.$$

We are finally in a position to prove the representation theorem.

Theorem I.4.3 Let $L(f)$ be a linear functional on $\mathcal{H}_p$ $(1 \le p \le 2)$ such that

I.4.8 $\qquad |L(f)| \le B\|f\|_{\mathcal{H}_p} \qquad \forall\ f \in \mathcal{H}_p.$

then there is a $\varphi \in \mathcal{H}_q$ $(q = \frac{p}{p-1})$ with $\|\varphi\|_{\mathcal{H}_q} \le \frac{2q}{q-1} B$ such that

I.4.9 $\qquad L(f) = \lim_{n \to \infty} E(f_n \varphi_n).$

Proof. We can consider $\mathcal{H}_p$ as the subset of $\mathcal{S}\mathcal{H}_p$ consisting of those sequences $\theta = (\theta_1, \theta_2, \ldots, \theta_n, \ldots)$ such that

I.4.10 $\qquad \theta_\nu = E(f|\mathcal{F}_\nu) - E(f|\mathcal{F}_{\nu-1}) \qquad \forall\ n$

for some $f \in \mathcal{H}_p$.

Thus $L(f)$ defines a linear functional on this subset. By the Hahn-Banach theorem, we can extend $L(f)$ to a linear functional $\Lambda(\theta)$ on $\mathcal{S}\mathcal{H}_p$ which has the same bound as $L(f)$. By Lemma I.4.1, we have a $\sigma \in \mathcal{S}\mathcal{H}_q$ with $\|\sigma\|_{\mathcal{S}\mathcal{H}_q} \le B$ such that

$$\Lambda(\theta) = \sum_{\nu=1}^{\infty} E(\sigma_\nu \theta_\nu)$$

Letting θ_ν be given by I.4.10 for $\nu \le n$ and be zero for $\nu > n$ we get

$$\Lambda(\theta) = L(f_n) = \sum_{\nu=1}^{n} E\left(\sigma_\nu \left[E(f|\mathcal{F}_\nu) - E(f|\mathcal{F}_{\nu-1})\right]\right)$$

$$= \sum_{\nu=1}^{n} E\left(f\left[E(\sigma_\nu|\mathcal{F}_\nu) - E(\sigma_\nu|\mathcal{F}_{\nu-1})\right]\right)$$

Now, letting

$$\varphi = \lim_{n\to\infty} \sum_{\nu=1}^{n} \left[E(\sigma_\nu|\mathcal{F}_\nu) - E(\sigma_\nu|f_{\nu-1})\right]$$

from Theorem I.4.1 we derive that $\varphi \in \mathcal{H}_q$,

$$\|\varphi\|_{\mathcal{H}_q} \le \frac{2q}{q-1} B$$

and

$$L(f_n) = E(f_n \varphi_n).$$

Now, since as $n \to \infty$

$$\|f - f_n\|_{\mathcal{H}_p} = \left[E\left(\left[\sum_{\nu=n+1}^{\infty} (\Delta f_\nu)^2 \right]^{p/2} \right) \right]^{1/p} \to 0$$

from the continuity of $L(f)$ I.4.9 must necessarily follow.

I.5 A remarkable representation for BMO functions.

The arguments of the last two sections can be combined to show that Theorem I.4.2 has a converse, namely:

<u>Theorem I.5.1</u>   Every $\varphi \in$ BMO <u>can be represented in the form</u>

$$\varphi = \sum_{\nu=1}^{\infty} \left[E(\sigma_\nu | \mathcal{F}_\nu) - E(\sigma_\nu | \mathcal{F}_{\nu-1}) \right]$$

<u>where</u>

$$\left[\sum_{\nu=1}^{\infty} \sigma_\nu^2 \right]^{1/2} \leq \sqrt{2} \ \|\varphi\|_{BMO}.$$

Indeed, we can show just as well that

<u>Theorem I.5.2</u>   Every $\varphi \in \mathcal{K}_q$ <u>can be represented in the form</u>

I.5.1 $$\varphi = \sum_{\nu=1}^{\infty} \left[E(\sigma_\nu | \mathcal{F}_\nu) - E(\sigma_\nu | \mathcal{F}_{\nu-1}) \right]$$

<u>where</u>

I.5.2
$$\left[E\left(\left[\sum_{\nu=1}^{\infty}\sigma_{\nu}^{2}\right]^{q/2}\right)\right]^{1/q} \le \sqrt{\frac{2}{p}}\,\|\varphi\|_{\mathcal{K}_q}\,.$$

Proof. Let $\varphi \in \mathcal{K}_q$ be fixed, and $p = \frac{q}{q-1}$. Then if $f \in \mathcal{H}_p$,

$$E(f_n\,\varphi_n) = \sum_{\nu=1}^{n} E(\Delta f_\nu\,\Delta\varphi_\nu)\,.$$

This identity was established in the course of the proof of Theorem I.3.1. Therefore, for all $m \ge n$ we have

$$E(f_m\,\varphi_m) - E(f_n\,\varphi_n) = \sum_{\nu=n+1}^{m} E(\Delta f_\nu\,\Delta\varphi_\nu) = E\bigl((f_m - f_n)\varphi_m\bigr)\,.$$

From I.3.3 we then get

$$\bigl|E(f_m\,\varphi_m) - E(f_n\,\varphi_n)\bigr| \le \sqrt{\frac{2}{p}}\,\|f_m - f_n\|_{\mathcal{H}_p}\,\|\varphi\|_{\mathcal{K}_q}\,.$$

Since $\|f_m - f_n\|_{\mathcal{H}_p} \to 0$ as $m, n \to \infty$, we must conclude that the sequence $E(f_n\,\varphi_n)$ is "Cauchy".

Set then for all $f \in \mathcal{H}_p$

$$L(f) = \lim_{n \to \infty} E(f_n\,\varphi_n)\,.$$

This given, from I.3.3 we derive

$$|L(f)| \le \sqrt{\frac{2}{p}} \; \|\varphi\|_{\mathcal{H}_q} \; \|f\|_{\mathcal{H}_p} .$$

This shows that $L(f)$ is a bounded linear functional on $\mathcal{H}_p$. The arguments of Theorem I.4.3 then yield the existence of a sequence of random variables $\{\sigma_n\}$, satisfying I.5.2, such that

$$L(f_n) = \sum_{\nu=1}^{n} E(\Delta f_\nu \; \sigma_\nu) = E\left(f_n \sum_{\nu=1}^{n} \left[E(\sigma_\nu|\mathcal{F}_\nu) - E(\sigma_\nu|\mathcal{F}_{\nu-1}) \right] \right).$$

But

$$L(f_n) = E(f_n \; \varphi_n) .$$

So we have

I.5.3 $\quad E(f_n \; \varphi_n) = E\left(f_n \sum_{\nu=1}^{n} \left[E(\sigma_\nu|\mathcal{F}_\nu) - E(\sigma_\nu|\mathcal{F}_{\nu-1}) \right] \right),$

for all $f \in \mathcal{H}_p$. In particular I.5.3 holds for all $f \in L_2$. This implies that

$$\varphi_n = \sum_{\nu=1}^{n} \left[E(\sigma_\nu | \mathcal{F}_\nu) - E(\sigma_\nu | \mathcal{F}_{\nu-1}) \right]$$

and I.5.1 necessarily follows.

Remark I.5.1   Although at this moment we cannot be sure that the product $f\varphi$ of an $f \in \mathcal{H}_p$ and a $\varphi \in \mathcal{K}_q$ ($q = \frac{p}{p-1}$) is integrable, we can yet express

$$\lim_{n \to \infty} E(f_n \varphi_n)$$

as an integral.

This can be seen as follows. If we go over the proof of Theorem I.3.1 we see that it yields just as well the inequality

I.5.4 $$\sum_{\nu=1}^{n} E\left(|\Delta f_\nu| \, |\Delta \varphi_\nu| \right) \le \sqrt{\frac{2}{p}} \, \|f\|_{\mathcal{H}_p} \, \|\varphi\|_{\mathcal{K}_q} .$$

Thus, passing to the limit as $n \to \infty$, we derive that whenever $f \in \mathcal{H}_p$ and $\varphi \in \mathcal{K}_q$ the function

$$\sum_{\nu=1}^{\infty} |\Delta f_\nu| \, |\Delta \varphi_\nu|$$

is integrable. We therefore conclude that

$$\lim_{n\to\infty} E(f_n \varphi_n) = E\left(\sum_{\nu=1}^{\infty} \Delta f_\nu \, \Delta\varphi_\nu \right).$$

,

II. INEQUALITIES BETWEEN THE SQUARE FUNCTION

AND

THE MAXIMAL FUNCTION.

II.1 <u>The equivalence of $\mathcal{H}_q$ and $\mathcal{K}_q$ norms</u>.

For an integrable function f we have set

$$f^* = \sup_n |E(f|\mathfrak{F}_n)|, \quad S(f) = \sqrt{\sum_{n=1}^{\infty} [\Delta f_n]^2}$$

where $\Delta f_n = E(f|\mathfrak{F}_n) - E(f|\mathfrak{F}_{n-1})$. It has become customary to call f^* and $S(f)$ the "maximal" and "square" functions respectively.

In this chapter we shall present proofs of the inequalities

II.1.1 $$\left[E(f^{*p})\right]^{1/p} \leq C'_p \left[E(S^p(f))\right]^{1/p},$$

II.1.2 $$\left[E(S^p(f))\right]^{1/p} \leq C''_p \left[E(f^{*p})\right]^{1/p},$$

for all $p \geq 1$. Here C'_p and C''_p denote constants depending only on p.

In particular we shall derive the remarkable fact that (when $f_0 = 0$) the norms $\|f\|_{\mathcal{H}_p}$ and $\left[E(f^{*p})\right]^{1/p}$ are equivalent.

As we shall see this result for $p = 1$ is equivalent to a very beautiful characterization of BMO functions.

There are several avenues open to us in establishing these inequalities. We shall follow here a path dictated by our desire to obtain reasonable expressions for C_p' and C_p''.

To this end we shall have to break up the proof of II 1.1 and II 1.2 into several cases according as $1 \leq p \leq 2$ or $2 \leq p < \infty$.

The easiest to obtain is II 1.2 for $p \geq 2$.

This we can derive immediately. We shall state it as follows

Theorem II 1.1 Let $2 \leq q < \infty$ then

II 1.3 $$\left[E\left(S^q(f)\right)\right]^{1/q} \leq \sqrt{2q} \left[E\left(f^{*q}\right)\right]^{1/q} .$$

Proof. We have

$$E\left(S_n^q(f)\right) = \sum_{\nu=1}^{n} E\left(S_\nu^q - S_{\nu-1}^q\right) \leq q/2 \sum_{\nu=1}^{n} E\left(S_\nu^{q-2}\left(S_\nu^2 - S_{\nu-1}^2\right)\right) .^{(*)}$$

Now, setting

(*) Here we have used the inequality $\rho^\alpha - 1 \leq \alpha(\rho-1) \rho^{\alpha-1}$

(valid when $\rho, \alpha \geq 1$) with $\alpha = q/2$, $\rho = \left(S_\nu/S_{\nu-1}\right)^2$.

II.1

$$\theta_\nu = s_\nu^{q-2} - s_{\nu-1}^{q-2}$$

we get

II 1.4
$$\begin{aligned} E\left(S_n^q(f)\right) &\leq q/2 \sum_{\nu=1}^{n} \sum_{\mu=1}^{\nu} E\left(\theta_\mu (s_\nu^2 - s_{\nu-1}^2)\right) = \\ &= q/2 \sum_{\mu=1}^{n} \sum_{\nu=\mu}^{n} E\left(\theta_\mu (s_\nu^2 - s_{\nu-1}^2)\right) = \\ &= q/2 \sum_{\mu=1}^{n} E\left(\theta_\mu (s_n^2 - s_{\mu-1}^2)\right) . \end{aligned}$$

Note that since θ_μ is $\mathcal{F}_\mu$-measurable

II 1.5
$$\begin{aligned} E\left(\theta_\mu (s_n^2 - s_{\mu-1}^2)\right) &= E\left(\theta_\mu E\left(s_n^2 - s_{\mu-1}^2 | \mathcal{F}_\mu\right)\right) = \\ &= E\left(\theta_\mu E\left(|f_n - f_{\mu-1}|^2 | \mathcal{F}_\mu\right)\right) . \end{aligned}$$

On the other hand we have $|f_n - f_{\mu-1}| \leq 2 f_n^*$, thus, combining these estimates:

$$E\left(S_n^q\right) \leq 2q \, E\left(S_n^{q-2} f_n^{*2}\right) .$$

Using Hölder's inequality here with exponents $\frac{q}{q-2}$ and $\frac{q}{2}$ we then get

$$E(S_n^q) \le 2q \left[E(S_n^q)\right]^{1-2/q} \left[E(f_n^{*q})\right]^{2/q}.$$

Since for $f^* \in L_q$ there is no question about the integrability of S_n^q, we can make the appropriate cancellation and arrive at

$$\left[E(S_n^q)\right]^{2/q} \le 2q \left[E(f_n^{*q})\right]^{2/q}.$$

The integrability of S^q as well as II 1.3 are thus immediately derived upon letting $n \to \infty$.

The proof we have just completed, by a slight modification, yields also the equivalence of the $\mathcal{H}_q$ and $\mathcal{K}_q$ norms for $q \ge 2$.

Indeed, if we know that

$$E\left(|f_n - f_{\mu-1}|^2 | \mathcal{F}_\mu\right) \le E\left(\gamma^2 | \mathcal{F}_\mu\right) \quad \forall \ \mu = 1, 2, \ldots, n$$

then, using this fact in II 1.5 from II 1.4 we obtain

$$E(S_n^q) \le \frac{q}{2} E(S_n^{q-2} \gamma^2).$$

So, by Hölder's inequality we can derive as well

$$\left[E(S_n^q)\right]^{1/q} \geq \sqrt{q/2} \left[E(\gamma^q)\right]^{1/q} .$$

and this of course implies

$$\|f\|_{\mathcal{H}_q} \leq \sqrt{q/2} \ \|f\|_{\mathcal{K}_q} .$$

The reverse inequality is almost immediate.

In fact, since

$$E\left(|f - f_{\mu-1}|^2 | \mathcal{F}_\mu\right) \leq E\left(S^2 | \mathcal{F}_\mu\right) ,$$

this means that S itself lies in the class Γ_f defined in Section I.2 .

We can thus state

<u>Theorem II 1.2</u> <u>For</u> $2 \leq q < \infty$ <u>the spaces</u> $\mathcal{H}_q$ <u>and</u> $\mathcal{K}_q$ <u>coincide and</u> $\forall \ f \in \mathcal{K}_q$

$$\|f\|_{\mathcal{K}_q} \leq \|f\|_{\mathcal{H}_q} \leq \sqrt{q/2} \ \|f\|_{\mathcal{K}_q} .$$

We see from Theorem II 1.1 that all we need to conclude that both these spaces coincide with L_q, is the inequality II 1.1 for $p \geq 2$. However, the latter is somewhat more difficult to obtain. Our program here is to establish first the inequalities II 1.1 and II 1.2 for $1 \leq p \leq 2$ then derive II 1.1 for $p \geq 2$ by a "Duality" argument.

To see how this comes about, note that we have for $f \in \mathcal{H}_p$, $\varphi \in \mathcal{H}_q$ ($q = p/p-1$)

$$E\left(f_n \varphi_n\right) = \sum_{\nu=1}^{n} E'\left(\Delta f_\nu \, \Delta \varphi_\nu\right).$$

So by Schwarz's inequality

$$\left|E(f_n \varphi_n)\right| \leq E\left(\sqrt{\sum_{\nu=1}^{n} \Delta f_\nu^2} \, \sqrt{\sum_{\nu=1}^{n} [\Delta \varphi_\nu]^2}\right) = E\left(S_n(f) \, S_n(\varphi)\right).$$

Hölder's inequality then gives

$$\left|E\left(f_n \varphi_n\right)\right| \leq \left[E\left([S_n(f)]^p\right)\right]^{1/p} \left[E\left([S_n(\varphi)]^q\right)\right]^{1/q}.$$

Now, if for this particular p we are in possession of II 1.2, we can get

$$|E(f_n \varphi_n)| \le c_p'' \left[E(f^{*p})\right]^{1/p} \left[E\left([S_n(\varphi)]^q\right)\right]^{1/q} .$$

So, using the inequality of Lemma I.4.2 we finally arrive at

$$|E(f_n \varphi_n)| \le c_p'' q \, \|f\|_p \, \|S_n(\varphi)\|_q .$$

This, of course, implies

$$\|\varphi_n\|_q \le c_p'' q \, \|S_n(\varphi)\|_q ,$$

and thus also

$$\|\varphi_n^*\|_q \le c_p'' pq \, \|S_n(\varphi)\|_q .$$

In other words, II 1.1 for a given $1 < p < \infty$ is implied by II 1.2 for the "conjugate" exponent. This might suggest that we should concentrate on showing II 1.2 and then get II 1.1 by "duality". Unfortunately, for two good reasons we cannot take this path.

First, we fail to see how this path could possibly lead to II 1.1 for $p = 1$. Secondly, even if we gave up this special case, the constant c_p' it leads to has the unpleasant feature that it tends to ∞ as $p \to 1$.

These objections do not apply when deriving II 1.1 for $2 \leq p < \infty$ from II 1.2 for $1 \leq p \leq 2$, since II 1.1 is false for $p = \infty$.

II.2 Proof of the inequalities for $1 \leq p \leq 2$.

The basic idea here will be similar to that used in [15] where we treat the special case $p = 1$. Namely, we shall show that II 1.1 and II 1.2 are both consequences of the inequality I.3.3.

Let us start with II 1.1.

A most natural thing to do in this case is to write

II 2.1
$$E\left(f_n^{*p}\right) = \sum_{\nu=1}^{n} E\left(|f_\nu|^p \chi_{E_\nu}\right)$$

where

II 2.2
$$E_\nu = \{f_{\nu-1}^* < f_n^* \; ; \; |f_\nu| = f_n^*\}$$

Now, if we set

$$\theta_\nu = |f_\nu|^{p-1} \operatorname{sign} f_\nu ,$$

we have

$$\sum_{\nu=1}^{n} E\left(|f_\nu|^p \chi_{E_\nu}\right) = \sum_{\nu=1}^{n} E\left(f_\nu \theta_\nu \chi_{E_\nu}\right) = \sum_{\nu=1}^{n} E\left(f_\nu \theta_\nu E\left(\chi_{E_\nu} | \mathfrak{F}_\nu\right)\right) .$$

Thus, II 2.1 can be written in the more suggestive form

$$E\left(f_n^{*p}\right) = E\left(f_n \, \varphi_n\right),$$

where

II 2.3
$$\varphi = \sum_{\nu=1}^{n} \theta_\nu \, E\left(\chi_{E_\nu} | \mathcal{F}_\nu\right).$$

But now, I.3.3 yields

II 2.4
$$E\left(f_n^{*p}\right) \leq \sqrt{\frac{2}{p}} \, \|S_n(f)\|_p \, \|\varphi\|_{\mathcal{H}_q}.$$

This leads to the study of the $\mathcal{H}_q$ norm of functions φ of the type II 2.3.

The outcome is the following very interesting result:

<u>Lemma II 2.1</u> Let $\{\theta_n\}$ and $\{\varepsilon_n\}$ be sequences of random variables such that

II 2.5
$$\mathcal{F}(\theta_n) \subset \mathcal{F}_n,$$

II 2.6
$$\sum_{n=1}^{\infty} |\varepsilon_n| \leq 1,$$

__then if__ $\theta^* = \sup_n |\theta_n|$ __is in__ L_q $2 \le q \le \infty$, __the function__

$$\varphi = \sum_{\nu=1}^{\infty} \theta_\nu \, E\left(\varepsilon_\nu | \mathcal{F}_\nu\right)$$

__is in__ $\mathcal{K}_q$ __and__

II 2.7 $$\|\varphi\|_{\mathcal{K}_q} \le \sqrt{5}\, p\, \|\theta^*\|_q .$$

Let us postpone the proof to the next section and proceed with our arguments. In the present case $\theta^* = f_n^{*(p-1)}$ so for $q = \frac{p}{p-1}$

$$\left[E(\theta^{*q})\right]^{1/q} = \left[E(f_n^{*p})\right]^{1/q} .$$

Combining this with II 2.7 and II 2.4 we get

$$E(f_n^{*p}) \le \sqrt{\frac{2}{p}}\, \|S_n(f)\|_p\, \sqrt{5}\, p\, \left[E(f_n^{*p})\right]^{1/q} .$$

Since the integrability of f_n^{*p} is not in doubt when $S(f) \in L_p$, we can make the cancellation and obtain

II 2.8 $$\|f_n^*\|_p \le \sqrt{10\,p}\ \|S_n(f)\|_p\ .$$

Let us now work on II 1.2.

Here we shall relate $S_n(f)$ to f_n by means of an identity which is apparently due to Doob, namely

II 2.9 $$S_n^2(f) = f_n^2 - 2\sum_{\nu=1}^{n} f_{\nu-1}\,\Delta f_\nu\ ,$$

To verify this note that, if $f_0 = 0$,

$$f_n^2 = \sum_{\nu=1}^{n}(f_\nu + f_{\nu-1})(f_\nu - f_{\nu-1}) = \sum_{\nu=1}^{n}(\Delta f_\nu + 2 f_{\nu-1})\,\Delta f_\nu\ .$$

This given, it is tempting to estimate $E\!\left(S_n^p(f)\right)$ when $1 \le p \le 2$ by Hölder's inequality in the following manner

II 2.10 $$E\!\left(S_n^p(f)\right) \le \left[E\!\left(\frac{S_n^2(f)}{A^{2/p}}\right)\right]^{p/2} \left[E\!\left(A^{\frac{2}{2-p}}\right)\right]^{1-p/2}\ .$$

Taking account of the fact that our goal is II 1.2 we are led to take

$$A^{\frac{2}{2-p}} = f_n^{*p}$$

which yields

$$A^{2/p} = f_n^{*(2-p)} \qquad (*)$$

We then have, by II 2.9

II 2.11 $\quad E\left(\dfrac{S_n^2(f)}{A^{2/p}}\right) = E\left(\dfrac{f_n^2}{f_n^{*(2-p)}}\right) - 2\sum_{\nu=1}^{n} E\left(\Delta f_\nu \, f_{\nu-1} \, \dfrac{1}{f_n^{*(2-p)}}\right) \leq$

$$\leq E\left(f_n^{*p}\right) + 2\,|Q|\,,$$

and we can write

$$Q = \sum_{\nu=1}^{n} E\left(\Delta f_\nu \, f_{\nu-1} \left[E\left(\dfrac{1}{f_n^{*(2-p)}}\,\Big|\,\mathfrak{F}_\nu\right) - E\left(\dfrac{1}{f_n^{*(2-p)}}\,\Big|\,\mathfrak{F}_{\nu-1}\right)\right]\right)$$

or better yet

(*) We can always reduce ourselves to cases in which $f_n^* > 0$ a.s.

$$Q = E(f_n \varphi_n)$$

where

II 2.12 $\quad \varphi = \sum_{\nu=1}^{n} f_{\nu-1} \left[E\left(\frac{1}{f_n^{*(2-p)}} \mid \mathfrak{F}_\nu \right) - E\left(\frac{1}{f_n^{*(2-p)}} \mid \mathfrak{F}_{\nu-1} \right) \right]$.

And now, we are again in the domain of I 3.3. If we use it and combine II 2.11 and II 2.10 we get

$$E\left(S_n^p(f)\right) \leq \left[E\left(f_n^{*p}\right) + 2\sqrt{\frac{2}{p}} \, \|S_n(f)\|_p \, \|\varphi\|_{\mathcal{H}_q} \right]^{p/2} \left[E\left(f_n^{*p}\right) \right]^{1-p/2}$$

It is convenient to set here $E(f_n^{*p}) = 1$ and raise the whole thing to the power $2/p$ to get

II 2.13 $\quad \|S_n(f)\|_p^2 \leq 1 + 2\sqrt{\frac{2}{p}} \, \|S_n(f)\|_p \, \|\varphi\|_{\mathcal{H}_q}$.

We are thus again left with having to estimate $\|\varphi\|_{\mathcal{H}_q}$. This is done by means of a further auxilary result:

Lemma II.2.2 <u>Let $\{\theta_n\}$ be a sequence of random variables such that $\mathfrak{J}(\theta_n) \subset \mathfrak{J}_n$, then if $\theta^* = \max |\theta_n|$ is in $L_p^{(*)}$ the function</u>

$$\varphi = \sum_{\nu=1}^{\infty} \theta_{\nu-1} \left[E\left(\frac{1}{\theta^{*(2-p)}} \Big| \mathfrak{J}_\nu\right) - E\left(\frac{1}{\theta^{*(2-p)}} \Big| \mathfrak{J}_{\nu-1}\right) \right]$$

<u>is in</u> $\mathcal{K}_q$ $\left(q = \frac{p}{p-1}\right)$ <u>and</u>

II 2.14 $\qquad \|\varphi\|_{\mathcal{K}_q} \leq \sqrt{2} \; \|\theta^*\|_p^{(p-1)} \; .$

We shall postpone the proof to the next section and proceed with our reasoning. In our case $\theta^* = f_n^*$ so, using II 2.14 in II 2.13, (in the case $E(f_n^{*p}) = 1$) we get

$$\|S_n(f)\|_p^2 \leq 1 + \frac{4}{\sqrt{p}} \; \|S_n(f)\|_p$$

or better

$$\|S_n(f)\|_p \leq \frac{2 + \sqrt{p+4}}{\sqrt{p}} \leq 5 \; .$$

So, in general we obtain

---

[*] For $p = 1$ this extra condition may be dropped.

II 2.15 $\qquad \|S_n(f)\|_p \leq 5 \|f_n^*\|_p \quad \forall \ 1 \leq p \leq 2$.

Going back to the "duality" argument at the end of Section II.1 we see that II 2.15 yields

II 2.16 $\qquad \|f_n^*\|_p \leq 5 \ pq \ \|S_n(f)\|_p \quad \forall \ 2 \leq p < \infty$.

II 3. Proofs of the auxiliary lemmas.

Let us start with Lemma II 2.1.

First of all note that there is no problem with convergence since

$$\sum_{\nu=1}^{\infty} E\left(|\theta_\nu| \ E\left(|\epsilon_\nu| \big| \mathcal{F}_\nu\right)\right) \leq E\left(\theta^*\right)$$

and supposedly θ^* is in L_q $(q \geq 2)$.

Also, it is not difficult to show that φ is L_q (but only when $q < \infty$). (*)

However, the estimate in II 2.7 is a little more delicate.

(*) Just integrate $\sum_{\nu=1}^{n} \theta_\nu \ E(\epsilon_\nu | \mathcal{F}_\nu)$ with a test function g in L_p and estimate the result by $E\left(\theta^* g^*\right)$, then use Lemma I.4.2.

To this end, note that

$$\varphi_{n-1} = \sum_{\nu=1}^{n-1} \theta_\nu E(\epsilon_\nu | \mathcal{F}_\nu) + E\left(\sum_{\nu=n}^{\infty} \theta_\nu \epsilon_\nu | \mathcal{F}_{n-1}\right).$$

Now, set for convenience

$$\Psi_n = \sum_{\nu=n}^{\infty} \theta_\nu E(\epsilon_\nu | \mathcal{F}_\nu).$$

This given we note that

$$\varphi - \varphi_{n-1} = \Psi_n - E(\Psi_n | \mathcal{F}_{n-1}).$$

Therefore

II 3.1
$$E\left(|\varphi - \varphi_{n-1}|^2 | \mathcal{F}_n\right) = E\left(\Psi_n^2 | \mathcal{F}_n\right) - 2E(\Psi_n | \mathcal{F}_n) E(\Psi_n | \mathcal{F}_{n-1}) + \left[E(\Psi_n | \mathcal{F}_{n-1})\right]^2.$$

We shall have to estimate each of these terms separately. We have, using II 2.5 and II 2.6

$$E(\Psi_n^2|\mathfrak{F}_n) \leq 2 \sum_{\nu=n}^{\infty} \sum_{\mu=\nu}^{\infty} E\left(|\theta_\nu| E(|\epsilon_\nu||\mathfrak{F}_\nu) |\theta_\mu| E(|\epsilon_\mu||\mathfrak{F}_\mu) |\mathfrak{F}_n\right).$$

$$= 2 \sum_{\nu=n}^{\infty} \sum_{\mu=\nu}^{\infty} E\left(|\theta_\nu| E(|\epsilon_\nu||\mathfrak{F}_\nu) |\theta_\mu| |\epsilon_\mu| |\mathfrak{F}_n\right)$$

$$\leq 2 \sum_{\nu=n}^{\infty} E\left(|\theta_\nu| E(|\epsilon_\nu||\mathfrak{F}_\nu) \theta^* |\mathfrak{F}_n\right)$$

$$= 2 \sum_{\nu=n}^{\infty} E\left(|\theta_\nu| |\epsilon_\nu| E(\theta^*|\mathfrak{F}_\nu) |\mathfrak{F}_n\right)$$

$$\leq 2 \sum_{\nu=n}^{\infty} E\left(\theta^* |\epsilon_\nu| \sup_{\nu \geq 1} E(\theta^*|\mathfrak{F}_\nu) |\mathfrak{F}_n\right)$$

$$\leq 2 E\left(\theta^* \sup_{\nu \geq 1} E(\theta^*|\mathfrak{F}_\nu) |\mathfrak{F}_n\right).$$

For the third term we have

$$|E(\Psi_n|\mathfrak{F}_{n-1})| \leq \sum_{\nu=n}^{\infty} E\left(|\theta_\nu| E(|\epsilon_\nu||\mathfrak{F}_\nu)|\mathfrak{F}_{n-1}\right) =$$

$$= \sum_{\nu=n}^{\infty} E\left(|\theta_\nu| |\epsilon_\nu| |\mathfrak{F}_{n-1}\right) \leq$$

$$\leq E(\theta^*|\mathfrak{F}_{n-1}) \leq \sup_{\nu \geq 1} E(\theta^*|\mathfrak{F}_\nu).$$

Quite similarly we get

$$|E(\Psi_n|\mathfrak{F}_n)| \leq E(\theta^*|\mathfrak{F}_n)$$

Note that since $\theta^* \in L_q$ and $\bigvee_{n=0}^{\infty} \mathfrak{F}_n = \mathfrak{F}$ we have

$$\theta^* \leq \sup_{\nu \geq 1} E\left(\theta^* | \mathfrak{F}_\nu\right) = \theta^{**}$$

These three estimates can thus be combined with II 3.1 to give

$$E\left(|\varphi - \varphi_{n-1}|^2 | \mathfrak{F}_n\right) \leq 2 E\left(\theta^{**2} | \mathfrak{F}_n\right) + 2 \theta^{**} E\left(\theta^* | \mathfrak{F}_n\right) + \left[\theta^{**}\right]^2.$$

Conditioning with respect to $\mathfrak{F}_n$ we finally get

$$E\left(|\varphi - \varphi_{n-1}|^2 | \mathfrak{F}_n\right) \leq E\left(\gamma^2 | \mathfrak{F}_n\right)$$

with

$$\gamma = \sqrt{5} \; \theta^{**}.$$

But, from Lemma I.4.2 we derive

$$\left[E\left(\gamma^q\right)\right]^{1/q} \leq \sqrt{5} \; p \left[E\left(\theta^{*q}\right)\right]^{1/q}.$$

This completes the proof of Lemma II 2.1.

II.3

Let us now work on Lemma II 2.2.

This is somewhat easier to prove. Indeed, if we set

$$g_\nu = E\left(\frac{1}{\theta^{*2-p}} \mid \mathcal{F}_\nu\right), \quad \Delta g_\nu = g_\nu - g_{\nu-1},$$

we have here

$$\varphi_n = \sum_{\nu=1}^{n} \theta_{\nu-1} \Delta g_\nu .$$

Thus, for $N \geq n$

II 3.2
$$E\left(|\varphi_N - \varphi_{n-1}|^2 \mid \mathcal{F}_n\right) = \sum_{\nu=n}^{N} E\left(\theta_{\nu-1}^2 \left[\Delta g_\nu\right]^2 \mid \mathcal{F}_n\right) \leq$$
$$\leq \sum_{\nu=n}^{N} E\left(\theta_{\nu-1}^{*2} \left[\Delta g_\nu\right]^2 \mid \mathcal{F}_n\right) ..$$

Where we have set as usual

$$\theta_\nu^* = \max_{0 \leq \mu \leq \nu} |\theta_\mu| .$$

But

$$\theta^*_{n-1} \Delta g_n \leq \theta^*_{n-1} E\left(\frac{1}{\theta^{*2-p}_n}\Big|\mathfrak{F}_n\right) \leq E\left(\theta^{*p-1}\Big|\mathfrak{F}_n\right)$$

and

$$\theta^*_{n-1} \Delta g_n \geq - \theta^*_{n-1} E\left(\frac{1}{\theta^{*2-p}_{n-1}}\Big|\mathfrak{F}_{n-1}\right) \geq - E\left(\theta^{*p-1}\Big|\mathfrak{F}_n\right),$$

So we must conclude that

II 3.3
$$\left[\theta^*_{n-1} \Delta g_n\right]^2 \leq E\left(\theta^{*2(p-1)}\Big|\mathfrak{F}_n\right).$$

Furthermore, for $\nu - 1 \geq n$, we have

$$E\left(\theta^{*2}_{\nu-1}\left[\Delta g_\nu\right]^2\Big|\mathfrak{F}_n\right) = E\left(\theta^{*2}_{\nu-1}\left[g^2_\nu - g^2_{\nu-1}\right]\Big|\mathfrak{F}_n\right).$$

This gives

II 3.4
$$\sum_{\nu=n+1}^{N} E\left(\theta^{*2}_{\nu-1}\left[\Delta g_\nu\right]^2\Big|\mathfrak{F}_n\right) =$$

$$= \sum_{\nu=n+1}^{N} E\left(\theta^{*2}_{\nu-1} g^2_\nu\Big|\mathfrak{F}_n\right) - \sum_{\nu=n+1}^{N} E\left(\theta^{*2}_{\nu-1} g^2_{\nu-1}\Big|\mathfrak{F}_n\right) \leq$$

$$\leq \sum_{\nu=n+1}^{N} E\left(\theta_\nu^{*2} g_\nu^2 | \mathfrak{F}_n\right) - \sum_{\nu=n+1}^{N} E\left(\theta_{\nu-1}^{*2} g_{\nu-1}^2 | \mathfrak{F}_n\right) \leq$$

$$\leq E\left(\theta_N^{*2} g_N^2 | \mathfrak{F}_n\right) .$$

But

$$\theta_N^* g_N = E\left(\frac{\theta_N^*}{\theta^{*2-p}} | \mathfrak{F}_N\right) \leq E\left(\theta^{*p-1} | \mathfrak{F}_N\right) ,$$

consequently, by Schwarz's inequality

$$\left(\theta_N^* g_N\right)^2 \leq E\left(\theta^{*2(p-1)} | \mathfrak{F}_N\right) ,$$

Substituting in II 3.4 and combining with II 3.3 and II 3.2 we finally get

II 3.5 $\qquad E\left(|\varphi_N - \varphi_{n-1}|^2 | \mathfrak{F}_n\right) \leq 2 E\left(\theta^{*2(p-1)} | \mathfrak{F}_n\right) .$

In particular, this implies that $\{\varphi_n\}$ is a Cauchy sequence in L_2. So at least $\varphi \in L_2$. Furthermore by letting $N \to \infty$ in II 3.5 we obtain that

$$E\left(|\varphi - \varphi_{n-1}|^2 | \mathfrak{F}_n\right) \leq E\left(\gamma^2 | \mathfrak{F}_n\right)$$

with

$$\gamma = \sqrt{2} \; \theta^{*(p-1)} \; .$$

This immediately yields II 2.14 and the proof of Lemma II 2.2 is now complete.

II 4. <u>A further characterization of BMO functions</u>.

In this section we shall present proof of the following result:

<u>Theorem II 4.1</u>. <u>Every</u> $\varphi \in$ BMO <u>can be written in the form</u>

II 4.1 $$\varphi = \Xi + \sum_{\nu=0}^{\infty} E(\varepsilon_\nu | \mathfrak{F}_\nu)$$

<u>where</u> Ξ <u>and</u> $\{\varepsilon_\nu\}$ <u>are random variables such that</u>

II 4.2 $$\left\| |\Xi| + \sum_{\nu=0}^{\infty} |\varepsilon_\nu| \right\|_\infty \leq \sqrt{2} \left(2 + \sqrt{5}\right) \|\varphi\|_{BMO} .$$

The representation II 4.1 for BMO functions is intimately related to the case $p = 1$ of the inequalities II 1.1 and II 1.2, namely

II 4.3 $$E(f_n^*) \leq \sqrt{10}\; E\bigl(S_n(f)\bigr),$$

II 4.4 $$E\bigl(S_n(f)\bigr) \leq \bigl(2 + \sqrt{5}\bigr) E(f_n^*).$$

To see how this comes about, note that our proof of II 4.3 hinged on Lemma II 2.1 which, in the special case $p=1$, can be stated as follows:

<u>Lemma II 4.1</u>   <u>Let</u> $\{\varepsilon_n\}$ <u>be a sequence of random variables such that</u>

$$\left\|\sum_{\nu=0}^{\infty} |\varepsilon_\nu|\right\|_\infty \leq B$$

<u>then the function</u>

$$\varphi = \sum_{\nu=0}^{\infty} E(\varepsilon_\nu | \mathcal{F}_\nu)$$

<u>is</u> BMO <u>and</u>

$$\|\varphi\|_{BMO} \leq \sqrt{5}\; B.$$

We shall prove Theorem II 4.1 by showing that the inequality II 4.4 is essentially equivalent to the converse of this lemma.

But before we can proceed with our arguments we need to dispose of a real variable lemma which will play here a role analogous to that played by Lemma I.4.1 in the proof of Theorem I.5.1.

To this end we let $\mathcal{SH}$ denote the Banach space of sequences of random variables

$$\theta = (\theta_0, \theta_1, \theta_2, \ldots, \theta_n, \ldots)$$

with norm

$$\|\theta\| = E(\theta^*)$$

where $\theta^* = \sup_{n \geq 0} |\theta_n|$.

For our purposes it is sufficient to consider $\mathcal{SH}$ as the completion of the sequences with "constant tail", that is, sequences which are of the form

II 4.5 $\theta = (\theta_0, \theta_1, \theta_2, \ldots, \theta_N, \gamma, \gamma, \gamma, \ldots)$

for some $N \geq 0$.

The following holds:

<u>Lemma II 4.2</u> <u>Let</u> $\Lambda(\theta)$ <u>be a linear functional on</u> $\mathcal{SH}$ <u>such that</u>

II 4.6 $$|\Lambda(\theta)| \leq c \, \|\theta\|$$

<u>then there are random variables</u>

$$\Xi, \, \varepsilon_0, \, \varepsilon_1, \, \varepsilon_2, \, \ldots, \, \varepsilon_n, \, \ldots$$

<u>with</u>

II 4.7 $$\left\| |\Xi| + \sum_{\nu=0}^{\infty} |\varepsilon_\nu| \right\|_\infty \leq c$$

<u>such that when</u> θ <u>is given by</u> II 4.5 <u>we have</u>

$$\Lambda(\theta) = \sum_{\nu=0}^{N} E(\varepsilon_\nu \, \theta_\nu) + E\left(\gamma\left[\Xi + \sum_{\nu=N+1}^{\infty} \varepsilon_\nu\right]\right).$$

Proof. Clearly when

$$\theta_n = \begin{cases} \theta_\nu & \text{for} \quad n = \nu \\ 0 & \text{for} \quad n \neq \nu \end{cases}$$

$\Lambda(\theta)$ is of the form

$$\Lambda(\theta) = E(\varepsilon_\nu \theta_\nu)$$

with $\varepsilon_\nu \in L_\infty$. For the same reason when

$$\theta = \begin{cases} 0 & \text{for } \nu \leq N, \\ \gamma & \text{for } \nu \geq N+1, \end{cases}$$

$\Lambda(\theta)$ is of the form

$$\Lambda(\theta) = E(\gamma \, \Xi_N)$$

with Ξ_N in L_∞.

This given, when θ is of the form II 4.5, by linearity we must have

II 4.8 $$\Lambda(\theta) = \sum_{\nu=0}^{N} E(\varepsilon_\nu \theta_\nu) + E(\gamma \, \Xi_N) \, .$$

Using this same relation with N replaced by $N+1$, but on the same sequence II 4.5, we immediately derive

II 4.9 $$\Xi_N = \varepsilon_N + \Xi_{N+1}.$$

Now, let us choose in II 4.5

$$\theta_\nu = \delta \text{ sign } \varepsilon_\nu \quad \text{for} \quad \nu = 0, 1, \ldots, N$$
$$\gamma = \delta \text{ sign } \Xi_N,$$

where δ is some non-negative integrable function.

The hypothesis II 4.6 then gives

$$\sum_{\nu=0}^{N} E\bigl(\delta|\varepsilon_\nu|\bigr) + E\bigl(\delta|\Xi_N|\bigr) \le c\, E(\delta).$$

This clearly implies

II 4.10 $$\sum_{\nu=0}^{N} |\varepsilon_\nu| + |\Xi_N| \le c \quad \text{a.s.}$$

Thus, in view of II 4.9 the sequence Ξ_N converges almost surely, furthermore the limit Ξ satisfies

$$\Xi_N = \Xi + \sum_{\nu=N+1}^{\infty} \varepsilon_\nu \qquad \forall\ N \geq 0.$$

But this relation together with II 4.8 and II 4.10 completes the proof of the lemma.

Let us proceed now with the proof of Theorem II 4.1.

Suppose the inequality

II 4.11 $$E\left(S_n(f)\right) \leq c\ E\left(f_n^*\right)$$

holds for all f with integrable f^*.

Then by Fefferman's inequality

$$\left|E\left(f_N \varphi\right)\right| \leq \sqrt{2}\ \|\varphi\|_{BMO}\ c\ E\left(f_N^*\right).$$

This means that for a fixed $\varphi \in BMO$, the functional

$$L_\varphi(f) = E(f\varphi)$$

has norm bounded by $\sqrt{2}\ c\ \|\varphi\|_{BMO}$ on the subset of $\mathcal{H}$ consisting of those sequences $\theta = (\theta_0, \theta_1, \ldots, \theta_n, \ldots)$ of the form

$$\theta_\nu = E(f_N | \mathcal{F}_\nu) \quad \nu = 0, 1, \ldots,$$

By the Hahn-Banach theorem, we can extend $L_\varphi(f)$ to all of $\mathcal{SH}$ without increasing the norm.

Lemma II 4.2 then yields random variables Ξ, $\{\epsilon_\nu\}$ satisfying

$$|\Xi| + \sum_{\nu=0}^\infty |\epsilon_\nu| \le \sqrt{2} \; c \; \|\varphi\|_{BMO} \, ,$$

for which we also have, for all f in L_1:

$$E(f_N \varphi) = \sum_{\nu=0}^N E(\epsilon_\nu \, f_\nu) + E\left(f_N \left[\Xi + \sum_{\nu=N+1}^\infty \epsilon_\nu\right]\right).$$

This of course gives

$$E(f_N \varphi) = E\left(f_N \left[\Xi + \sum_{\nu=0}^\infty E(\epsilon_\nu | \mathcal{F}_\nu)\right]\right).$$

The arbitrariness of f then yields

$$\varphi_N = E\left(\Xi + \sum_{\nu=0}^\infty E(\epsilon_\nu | \mathcal{F}_\nu) \;\Big|\; \mathcal{F}_N\right).$$

Letting $N \to \infty$ the representation II 4.1 immediately follows.

The constant in II 4.2 is then due to the fact that we can take $c = 2 + \sqrt{5}$ in II 4.11.

<u>Remark II 4.1</u> It is interesting to note that viceversa, if we know there is a universal constant c_1 such that every BMO function φ can be written in the form II 4.1 with

$$\text{II 4.12} \qquad |\Xi| + \sum_{\nu=0}^{\infty} |\varepsilon_\nu| \leq c_1 \, \|\varphi\|_{BMO} \, ,$$

then we can show that II 4.11 must necessarily follow with $c = 2c_1$.

We give only a brief sketch of the argument here since we have already treated this matter in [15].

The idea is that for a given $f \in \mathcal{H}_1$ we can write

$$E\bigl(S(f)\bigr) = \sum_{\nu=1}^{\infty} E\left(\frac{[\Delta f_\nu]^2}{S(f)}\right) = E(f\varphi)$$

where

$$\varphi = \sum_{\nu=1}^{\infty} \left[E\left(\frac{\Delta f_\nu}{S(f)} \,\Big|\, \mathcal{F}_\nu\right) - E\left(\frac{\Delta f_\nu}{S(f)} \,\Big|\, \mathcal{F}_{\nu-1}\right) \right] .$$

This given, since

$$\sum_{\nu=1}^{\infty} \left[\frac{\Delta f_\nu}{S(f)} \right]^2 = 1 ,$$

from theorem I 4.1 (with $q = \infty$) we get

II 4.13 $$\|\varphi\|_{BMO} \leq 2$$

Using the representation II 4.1 we then obtain

$$E\bigl(S(f)\bigr) = E(f\,\Xi) + \sum_{\nu=0}^{\infty} E\bigl(f_\nu\, e_\nu\bigr) \leq$$

$$\leq E\left(f^* \left[|\Xi| + \sum_{\nu=0}^{\infty} |e_\nu| \right] \right) .$$

Finally, combining II 4.13 with II 4.12 we get

$$E\bigl(S(f)\bigr) \leq 2c_1\, E(f^*)$$

as asserted.

II 5 Square function inequalities for supermartingales and Burkholder's weak-L_1 inequality.

In [3] Burkholder established indirectly the existence of a universal constant c such that

II 5.1 $$P[S(f) \geq \lambda] \leq \frac{c}{\lambda} E(|f|) \, ,$$

$\forall \; f \in L_1$ and all $\lambda > 0$.

Later Gundy [17] discovered a remarkable decomposition for L_1-bounded martingales which led to a direct proof of II 5.1.

In this section we shall show that Doob's identity

II 5.2 $$f_0^2 + \sum_{\nu=1}^{n} (f_\nu - f_{\nu-1})^2 + 2 \sum_{\nu=1}^{n} f_{\nu-1}(f_\nu - f_{\nu-1}) = f_n^2 \, ,$$

which was the point of departure in our proof of the Burgess Davis inequality II 4.4, yields also a remarkably simple proof of II 5.1 with $c = 6$.

Before carrying on with our arguments we want to mention that Burkholder in [2], starting with more or less the same idea, but with a little more effort, establishes II 5.1 with $c = 3$. Apparently, the best constant lies somewhere between 2 and $2\sqrt{2}$.

Let us note first that II 5.2 is purely algebraic, thus if $\{Q_\nu\}$ is any sequence of random variables and we set

$$S^2(Q_n) = Q_0^2 + \sum_{\nu=1}^{n} (Q_\nu - Q_{\nu-1})^2$$

we have

$$E(S^2(Q_n)) = E(Q_n^2) + 2 \sum_{\nu=1}^{n} E(Q_{\nu-1}(Q_{\nu-1} - Q_\nu)) \ .$$

Thus, if $\{Q_\nu\}$ is adapted to $\{\mathcal{F}_\nu\}$,[(*)] we also have

II 5.3 $\quad E(S^2(Q_n)) = E(Q_n^2) + 2 \sum_{\nu=1}^{n} E(Q_{\nu-1} E(Q_{\nu-1} - Q_\nu | \mathcal{F}_{\nu-1})) \ .$

This given, to prove II 5.1, we first assume $f \geq 0$ and set in II 5.3 $Q_\nu = f_\nu \wedge \lambda$. Since this sequence is a supermartingale, $E(Q_{\nu-1} - Q_\nu | \mathcal{F}_{\nu-1}) \geq 0$, and II 5.3 yields

$$E(S^2(Q_n)) \leq \lambda E(Q_n) + 2\lambda \sum_{\nu=1}^{n} E(Q_{\nu-1} - Q_\nu)$$

[(*)] I.e. $\mathcal{F}(Q_\nu) \subset \mathcal{F}_\nu$, $\forall \nu$

or, better yet

II 5.4 $$E\left(S^2(Q_n)\right) \leq 2\lambda E(Q_0) \leq 2\lambda E(f).$$

But now note that when $f^* \leq \lambda$ Q_ν and f_ν are identical. Thus II 5.4 implies

II 5.5 $$P\left[S_n(f) \geq \lambda;\; f^* \leq \lambda\right] \leq \frac{2}{\lambda} E(f).$$

Recalling the classical inequality

II 5.6 $$P\left[f^* > \lambda\right] \leq \frac{1}{\lambda} E(|f|), \quad (*)$$

we immediately derive

$$P\left[S_n(f) \geq \lambda\right] \leq P\left[f^* > \lambda\right] + P\left[S_n(f) \geq \lambda;\; f^* \leq \lambda\right] \leq \frac{3}{\lambda} E(f).$$

The sign restriction on f is removed by means of the usual decomposition $f = f^+ - f^-$. Indeed, we then have $S_n(f) \leq S_n(f^+) + S_n(f^-)$,

---

(*) For this as well as Lemma I 4.2 a proof will be found in the next chapter.

thus

$$P\left[S_n(f) \geq \lambda\right] \leq P\left[S_n(f^+) \geq \frac{\lambda}{2}\right] + P\left[S_n(f^-) \geq \frac{\lambda}{2}\right] \leq$$

$$\leq \frac{6}{\lambda} E(f^+) + \frac{6}{\lambda} E(f^-) ,$$

and II 5.1 follows with c = 6 as asserted.

Remark II 5.1 It is to be noted that II 5.4 yields also

$$E\left(S_n^2(f) \chi(f^* \leq \lambda)\right) \leq 2 \lambda E(f) .$$

This implies the remarkable fact discovered by Austin [1] that the square function of an L_1-bounded martingale $\{f_n\}$ is square integrable over any set where f^* is bounded above.

We shall see in the next chapter that $S_n^2(f)$ is actually <u>exponentially</u> integrable over the same set!

Remark II 5.2 Going back to II 5.3, we see that if $\{Q_\nu\}$ is a martingale or a non-negative submartingale we necessarily have

$$E\left(S^2(Q_n)\right) \leq E\left(Q_n^2\right) ,$$

the remarkable fact is that an estimate on $E\left(S^2(Q_n)\right)$ can be obtained even when Q_ν is a non-negative <u>supermartingale</u>, provided of course Q_ν is dominated by a martingale. We have essentially used this idea in our proof of II 5.1 .

To see this assume that $E(Q_{\nu-1} - Q_\nu | \mathcal{F}_{\nu-1}) \geq 0$ and

$$0 \leq Q_\nu \leq E(\gamma | \mathcal{F}_\nu) \quad \text{for} \quad \nu = 0, 1, \ldots, n .$$

Then II 5.3 yields

$$E\left(S^2(Q_n)\right) \leq E\left(\gamma Q_n\right) + 2 \sum_{\nu=1}^{n} E\left(\gamma \, E(Q_{\nu-1} - Q_\nu | \mathcal{F}_{\nu-1})\right)$$

or better

II 5.7 $$E\left(S^2(Q_n)\right) \leq 2 \, E(\gamma A_n)$$

where

$$A_n = Q_n + \sum_{\nu=1}^{n} E\left(Q_{\nu-1} - Q_\nu | \mathcal{F}_{\nu-1}\right) .$$

Now, setting $Q_\nu - Q_{\nu-1} = \Delta Q_\nu$ we have

II.5

$$A_n = Q_0 + \sum_{\nu=1}^{n} \left[Q_\nu - E(Q_\nu | \mathcal{F}_{\nu-1}) \right] = Q_0 + \sum_{\nu=1}^{n} \left[\Delta Q_\nu - E(\Delta Q_\nu | \mathcal{F}_{\nu-1}) \right] .$$

Thus

$$E(A_n^2) \leq E(Q_0^2) + \sum_{\nu=1}^{n} E\left([\Delta Q_\nu]^2\right) = E\left(S^2(Q_n)\right) .$$

So, a use of Schwarz's inequality in II 5.7 finally yields

II 5.8 $$E\left(S^2(Q_n)\right) \leq 4 \, E(\gamma^2) .$$

III. BMO SEQUENCES AND POTENTIALS.

III.1 Exponential bounds for monotone sequences.

The condition

$$\sup_n \left\| \sqrt{E(|\varphi - \varphi_{n-1}|^2 | \mathcal{F}_n)} \right\|_\infty < \infty$$

obviously implies

$$\sup_n \left\| E(|\varphi - \varphi_{n-1}| | \mathcal{F}_n) \right\|_\infty < \infty .$$

The remarkable fact is that the implication also goes the other way around. Indeed, it can be shown that the norms

III 1.1 $\quad \|\varphi\|_{BMO_p} = \sup_n \left\| \left[E([\varphi - \varphi_{n-1}]^p | \mathcal{F}_n) \right]^{1/p} \right\|_\infty \quad (p \geq 1)$

are all equivalent.

This was first proved by John and Nirenberg [21] in the special case that the probability space is the d-dimensional unit cube I_0 and $\{\mathcal{F}_n\}$ is the sequence of finite fields obtained by successive dyadic partitions of I_0.

This result was actually only a by-product of the following theorem, which here and after will be referred to as the "John-Nirenberg theorem".

III.1   65

Theorem III 1.1   If

III 1.2 $$\|\varphi\|_{BMO_1} = \sup_n \|E(|\varphi - \varphi_{n-1}| \,|\, \mathcal{F}_n)\|_\infty \leq 1$$

then

III 1.3 $$P\left[|\varphi - \varphi_{n-1}| > \lambda \,|\, \mathcal{F}_n\right] \leq c_1 e^{-\lambda/c_2}$$

where c_1 and c_2 are universal constants.

To be sure, John and Nirenberg established this only in the special case of dyadic partitions of the unit cube. The above formulation is due to Gundy[*] and in this form it was proved more or less independently by several authors including Gundy, Herz [19], Fefferman and myself.

In this chapter we shall present a simple proof of III 1.3 with $c_1 = 2$ and $c_2 = 16$. Indeed, we shall show here that the following inequality holds in full generality

III 1.4 $$E\left(e^{|\varphi - \varphi_{n-1}|} \,|\, \mathcal{F}_n\right) \leq \frac{1}{1 - 8\|\varphi\|_{BMO_1}} .$$

---

[*] Personal communication.

In addition we shall also investigate some of the ramifications that the proof of this result leads to.

It should be pointed out that Theorem III 1.1 which is now a "classic", as well as the notion of a "BMO" function are fraught with important applications.

In addition to the original use made by John in [20], the reader is urged to have a look at the role III 1.3 plays in the incredibly beautiful derivation of the Harnack inequalities made by Moser in [25] and [26].

Our proof here will be of such a nature that it can be carried out entirely outside the "martingale" setting. To this end, for convenience, let us say that a sequence of random variables $\{A_n\}$ is "BMO(B)" if and only if the following conditions are satisfied

III 1.5

a) $\mathfrak{J}(A_n) \subset \mathfrak{J}_n$

b) $E(|A_n - A_{\nu-1}| \,|\, \mathfrak{J}_\nu) \leq B$ a.s. $\forall\, n \geq \nu \geq 1$.

This given, we can immediately prove the following remarkable result:

<u>Theorem III 1.2</u>   <u>If</u> $\{A_n\}$ <u>is</u> BMO(B) $(B < 1)$ <u>and non-decreasing, then for all</u> $n \geq \nu \geq 1$

III 1.6
$$E\left(e^{A_n - A_{\nu-1}} \,\big|\, \mathfrak{J}_\nu\right) \leq \frac{1}{1-B}.$$

Proof. First of all note that we need only prove III 1.6 for $\nu = 1$ and $A_0 = 0$. Indeed, by appropriately relabeling the sequence $\{\mathcal{F}_n\}$ and subtracting A_0 we can always reduce ourselves to that case.

Furthermore, it is easy to see that the condition III 1.5 implies $|A_\nu - A_{\nu-1}| \leq B \quad \forall \nu$ thus all the differences $A_\nu - A_{\nu-1}$ are necessarily bounded.

To prove III 1.6, let us proceed a little more generally and let $m(t)$ be non-decreasing in $[0, \infty]$. We then have

$$E\left(\int_0^{A_n} m(t)\,dt \Big| \mathcal{F}_1\right) = \sum_{\nu=1}^n E\left(\int_{A_{\nu-1}}^{A_\nu} m(t)\,dt \Big| \mathcal{F}_1\right) \leq$$

$$\leq \sum_{\nu=1}^n E\left(m(A_\nu)(A_\nu - A_{\nu-1}) \Big| \mathcal{F}_1\right).$$

Thus, setting $p_1 = m(A_1)$ and

$$p_\nu = m(A_\nu) - m(A_{\nu-1}) \quad \nu = 2, \ldots, n$$

we obtain

$$E\left(\int_0^{A_n} m(t)\,dt \Big| \mathcal{F}_1\right) \le \sum_{\nu=1}^n \sum_{\mu=1}^\nu E\left(p_\mu(A_\nu - A_{\nu-1}) \Big| \mathcal{F}_1\right) =$$

$$= \sum_{\mu=1}^n E\left(p_\mu(A_n - A_{\mu-1}) \Big| \mathcal{F}_1\right) =$$

$$= \sum_{\mu=1}^n E\left(p_\mu\, E(A_n - A_{\mu-1} | \mathcal{F}_\mu) \Big| \mathcal{F}_1\right) \le$$

$$\le B\, E\left(\sum_{\mu=1}^n p_\mu \Big| \mathcal{F}_1\right) = B\, E\left(m(A_n) \Big| \mathcal{F}_1\right).$$

Now, if we chose $m(t) = e^t$ this yields

$$E\left(e^{A_n} - 1 \Big| \mathcal{F}_1\right) \le B\, E\left(e^{A_n} \Big| \mathcal{F}_1\right)$$

and thus III 1.6 follows since the integrability of e^{A_n} is not in question.

Note, if we set $m(t) = t$ then we get

$$E\left(\frac{A_n^2}{2} \Big| \mathcal{F}_1\right) \le B\, E(A_n | \mathcal{F}_1) \le B^2$$

and thus also

$$E\left(|A_n - A_{\nu-1}|^2 \Big| \mathcal{F}_\nu\right) \le 2\, B^2 \;.$$

This should give at least an indication of the reasons behind the equivalence of the norms in III 1.1.

Remark III 1.1 If we go back to the notation introduced in Section I.1 we see that

$$E\left(S_n^2(\varphi) - S_{\nu-1}^2(\varphi)\,|\,\mathfrak{F}_\nu\right) \leq \|\varphi\|_{BMO}^2 \;.$$

In other words $\varphi \in BMO$ implies that the sequence $\{S_n^2(\varphi)\}$ is BMO under the present definition. We thus immediately obtain the following remarkable corollary

Theorem III 1.3 If $\|\varphi\|_{BMO_2} < 1$ then $e^{S^2(\varphi)}$ is integrable and

$$E\left(e^{S^2(\varphi)}\right) \leq \frac{1}{1 - \|\varphi\|_{BMO_2}^2} \;.$$

This result was also discovered by several authors including Burkholder [2] and Meyer [24].

Remark III 1.2 Let $\{f_n\}$ be a non-negative martingale and set

$$Q_n = \frac{f_n}{f_n^*} \quad (f_n^* = \max_{0 \leq \nu \leq n} f_\nu)$$

We can show that the sequence

$$A_n = \sum_{\nu=1}^{n} [\Delta Q_\nu]^2 \quad (\Delta Q_\nu = Q_\nu - Q_{\nu-1})$$

is BMO(3). Indeed, by Doob's identity

$$A_n - A_{\nu-1} = Q_n^2 - Q_{\nu-1}^2 + 2 \sum_{\mu=\nu}^{n} Q_{\mu-1} (Q_{\mu-1} - Q_\mu) .$$

Since $\{Q_n\}$ is a supermartingale bounded by 1 we immediately derive

$$E(A_n - A_{\nu-1} | \mathcal{F}_\nu) \leq E(Q_n^2 | \mathcal{F}_\nu) + Q_{\nu-1}^2 + 2 \sum_{\mu=\nu+1}^{n} E(Q_{\mu-1} E(Q_{\mu-1} - Q_\mu | \mathcal{F}_{\mu-1}) | \mathcal{F}_\nu)$$

$$\leq E(Q_n | \mathcal{F}_\nu) + 1 + 2 \sum_{\mu=\nu+1}^{n} E(Q_{\mu-1} - Q_\mu | \mathcal{F}_\nu) .$$

$$\leq 1 + 2Q_\nu \leq 3 .$$

Now, $f_n = f_n^* Q_n$ and taking differences

III 1.7 $\quad \Delta f_n = \left(f_n^* - f_{n-1}^* \right) Q_n + f_{n-1}^* \Delta Q_n = f_n^* - f_{n-1}^* + f_{n-1}^* \Delta Q_n$.

So we get

$$[\Delta f_n]^2 \leq 2\left(f_n^* - f_{n-1}^*\right)^2 + 2 f_{n-1}^{*2} [\Delta Q_n]^2,$$

$$\leq 2 f_n^*\left(f_n^* - f_{n-1}^*\right) + 2 f_n^* f^* [\Delta Q_n]^2,$$

therefore

$$\frac{1}{f^*} \sum_{n=1}^{\infty} \frac{[\Delta f_n]^2}{f_n^*} \leq 2 + 2 \sum_{n=1}^{\infty} [\Delta Q_n]^2$$

We thus have established the following surprising fact:

<u>Theorem III 1.4</u> <u>If $\{f_n\}$ is any non-negative martingale then</u>
$\forall \; \alpha < 1/6$

$$E\left(e^{\frac{\alpha}{f^*} \sum_{n=1}^{\infty} \frac{[\Delta f_n]^2}{f_n^*}}\right) \leq \frac{e^{2\alpha}}{1 - 6\alpha} \; .$$

In particular, we see that

$$E\left(e^{\frac{1}{12} \frac{S^2(f)}{\lambda^2}} \chi(f^* \leq \lambda)\right) \leq 2 e^{1/6} \; .$$

a result we had announced in the Remark II 5.1.

Remark III 1.3   Incidentally, the decomposition in III 1.7 yields a surprisingly simple proof of Burkholder's convergence theorem for martingale transforms, namely

Theorem III 1.5   Let $\{f_n\}$ be an L_1-bounded martingale and let $\{\lambda_n\}$ be a sequence of random variables such that

III 1.8
a) $\mathcal{F}(\lambda_n) \subset \mathcal{F}_n$
b) $\sup |\lambda_n| < \infty$ a.s.

Then the martingale

$$g_n = \sum_{\nu=1}^{n} \lambda_{\nu-1} \Delta f_\nu$$

converges a.s.

Proof. There is no loss in assuming $f_n \geq 0$ since every L_1-bounded martingale can be decomposed into the difference of two non-negative martingales.

III.1   73

This given, using III 1.7 we write

$$\Delta f_\nu = f^*_\nu - f^*_{\nu-1} + f^*_{\nu-1} E(\Delta Q_\nu | \mathcal{F}_{\nu-1}) + f^*_{\nu-1} [\Delta Q_\nu - E(\Delta Q_\nu | \mathcal{F}_{\nu-1})],$$

or better

III 1.9 $\quad \Delta f_\nu = f^*_\nu - f^*_{\nu-1} - f^*_{\nu-1} E(Q_{\nu-1} - Q_\nu | \mathcal{F}_{\nu-1}) + f^*_{\nu-1} \Delta h_\nu$

where we have set

$$h = \sum_{\nu=1}^{\infty} \left[\Delta Q_\nu - E(\Delta Q_\nu | \mathcal{F}_{\nu-1}) \right].$$

Incidentally, note that this function is BMO. Indeed

$$-1 \leq -E(Q_\nu | \mathcal{F}_{\nu-1}) \leq \Delta h_\nu = Q_\nu - E(Q_\nu | \mathcal{F}_{\nu-1}) \leq Q_\nu \leq 1$$

and thus (as we have seen in our previous remark)

$$E\left(|h_n - h_{\nu-1}|^2 | \mathcal{F}_\nu\right) \leq [\Delta h_\nu]^2 + \sum_{\mu=\nu+1}^{n} E([\Delta Q_\mu]^2 | \mathcal{F}_\nu) \leq 1 + 3.$$

Our decomposition III 1.9 then gives

$$g_n = \sum_{\nu=1}^{n} \lambda_{\nu-1} (f_\nu^* - f_{\nu-1}^*) - \sum_{\nu=1}^{n} \lambda_{\nu-1} f_{\nu-1}^* E\left(Q_{\nu-1} - Q_\nu | \mathcal{F}_{\nu-1}\right) +$$
$$+ \sum_{\nu=1}^{n} \lambda_{\nu-1} f_{\nu-1}^* \Delta h_\nu .$$

Now, the first series under our assumptions converges in a trivial way. The second series never gives any problem since

$$E\left(\sum_{\nu=1}^{\infty} E(Q_{\nu-1} - Q_\nu | \mathcal{F}_{\nu-1})\right) \leq E(Q_0) \leq 1.$$

Finally, the third series is well behaved or any set of the form $\{\sup_n f_n^* |\lambda_n| \leq c\}$ since on such a set it agrees identically with an L_2-bounded martingale (actually a BMO martingale!).

III.2 Proof of the John-Nirenberg theorem.

We shall obtain the inequality III 1.4 and therefore also the John-Nirenberg theorem from the following remarkable result.

<u>Theorem III 2.1</u> <u>If</u> $\{A_n\}$ <u>is</u> BMO(B) <u>then</u> $A_n^* = \max\limits_{0 \le \nu \le n} |A_\nu|$ <u>is</u> BMO(8B).

Proof. Let us set for convenience $B = 1$ and assume first that $A_0 = 0$. We shall proceed to show that $\forall$ n

$$E(A_n^* | \mathcal{F}_1) \le 8 .$$

To this end, for a given $\alpha > 0$ set

III 2.1 $$A_n^\alpha = \sum_{\mu=1}^n \chi(A_{\mu-1}^* \le \alpha;\ A_\mu^* > \alpha) A_\mu + \chi(A_n^* \le \alpha) A_n$$

Clearly, on the set $\{A_n^* > \alpha\}$

$$\sum_{\mu=1}^n \chi(A_{\mu-1}^* \le \alpha;\ A_\mu^* > \alpha) = 1 .$$

and therefore, using the fact that

$$|A_\nu - A_{\nu-1}| \leq B = 1$$

we have there

$$\alpha \leq |A_n^\alpha| \leq \alpha + 1 .$$

This means that whenever $\beta > \alpha$

III 2.2 $\quad E\left(|A_n^\beta - A_n^\alpha|\chi(A_n^* > \beta)\,|\mathcal{F}_1\right) \geq (\beta - \alpha - 1)\, P[A_n^* > \beta\,|\mathcal{F}_1] .$

On the other hand

$$|A_n^\beta - A_n^\alpha| \leq |A_n - A_n^\alpha| + |A_n - A_n^\beta| ,$$

and thus

$$E\left(|A_n^\beta - A_n^\alpha|\,|\mathcal{F}_1\right) \leq \sum_{\mu=1}^n E\left(\chi(A_{\mu-1}^* \leq \alpha;\, A_\mu^* > \alpha)\,|A_n - A_\mu|\,|\mathcal{F}_1\right) +$$

$$+ \sum_{\mu=1}^n E\left(\chi(A_{\mu-1}^* \leq \beta;\, A_\mu^* > \beta)\,|A_n - A_\mu|\,|\mathcal{F}_1\right) .$$

The assumption that $\{A_\nu\}$ is BMO(B) immediately gives then

$$E\left(|A_n^\beta - A_n^\alpha| \,|\, \mathcal{F}_1\right) \le B\, P\left[A_n^* > \alpha | \mathcal{F}_1\right] + B\, P\left[A_n^* > \beta | \mathcal{F}_1\right].$$

Combining with III 2.2 and $B = 1$ we derive

III 2.3 $\qquad (\beta - \alpha - 2)\, P\left[A_n^* > \beta | \mathcal{F}_1\right] \le P\left[A_n^* > \alpha | \mathcal{F}_1\right]$

setting $\beta = \alpha + 4$ and integrating with respect to α from 0 to ∞ we easily get

$$2\, E\left((A_n^* - 4)^+ | \mathcal{F}_1\right) \le E\left(A_n^* | \mathcal{F}_1\right).$$

This gives

$$E\left(A_n^* | \mathcal{F}_1\right) \le E\left((A_n^* - 4)^+ | \mathcal{F}_1\right) + 4 \le$$

$$\le \tfrac{1}{2}\, E\left(A_n^* | \mathcal{F}_1\right) + 4$$

or finally

$$E(A_n^* | \mathcal{F}_1) \le 8$$

as asserted.

To complete the proof, we see that in general we have (upon subtracting A_0)

$$E(\max_{1 \leq m \leq n} |A_m - A_0| \,|\, \mathfrak{F}_1) \leq 8B .$$

By a suitable relabeling of the sequence $\{\mathfrak{F}_\nu\}$ this implies

$$E(\max_{\nu \leq m \leq n} |A_m - A_{\nu-1}| \,|\, \mathfrak{F}_\nu) \leq 8B .$$

However, we see that

$$A_n^* - A_{\nu-1}^* \leq \max_{\nu \leq m \leq n} |A_m - A_{\nu-1}| .$$

That finishes our argument.

To obtain the inequality III 1.4 note that theorems III 1.2 and III 2.1 combined have the following immediate corollary

Theorem III 2.2 If $\{A_n\}$ is BMO(B) then

III 2.4
$$E\left(e^{\max_{\nu \leq m \leq n} |A_m - A_{\nu-1}|} \Big| \mathcal{F}_\nu\right) \leq \frac{1}{1 - 8B}.$$

Proof. We first derive, using III 1.6, when $A_0 = 0$:

$$E\left(e^{A_n^*} \Big| \mathcal{F}_1\right) \leq \frac{1}{1 - 8B}.$$

then III 2.4 immediately follows upon relabelling the fields $\{\mathcal{F}_\nu\}$ and subtracting $A_{\nu-1}$.

Note that III 2.4 implies that when

$$\sup_\nu \|E(|\varphi - \varphi_{\nu-1}| \, | \mathcal{F}_\nu)\|_\infty \leq B$$

then

$$E\left(e^{\sup_{n \geq \nu} |\varphi_n - \varphi_{\nu-1}|} \Big| \mathcal{F}_\nu\right) \leq \frac{1}{1 - 8B}$$

a-fortiori III 1.4 must hold as well.

Finally we note that as a by-product of the inequality

$$E\left(|A_n - A_{\nu-1}|^2 \Big| \mathcal{F}_\nu\right) \leq 2 B^2$$

obtained for monotone BMO(B) sequences we derive

$$E\left((A_n^* - A_{\nu-1}^*)^2 | \mathcal{F}_\nu\right) \leq 128\, B^2$$

for all BMO(B) sequences. This yields, in the same manner III 2.4 is obtained from III 1.6:

$$E\left(\max_{\nu \leq m \leq n} |A_m - A_{\nu-1}|^2 | \mathcal{F}_\nu\right) \leq 128\, B^2 .$$

Thus one of the by-products of our efforts here is the following result

<u>Theorem III 2.3</u> <u>If</u> φ <u>is such that</u>

$$\sup_{n \geq \nu} \left\| E\left(|\varphi_n - \varphi_{\nu-1}| | \mathcal{F}_\nu\right) \right\|_\infty \leq B$$

<u>then</u> φ <u>is</u> BMO <u>and</u>

$$\|\varphi\|_{BMO} \leq 8\sqrt{2}\, B.$$

Remark III 2.1   C. Herz [19] has shown that in general the norms

$$\text{III 2.5} \qquad \|\varphi\|_{BMO_p^+} = \sup_\nu \left\| \left[E([\varphi - \varphi_\nu]^p | \mathcal{F}_\nu) \right]^{1/p} \right\|_\infty$$

unlike those defined in III 1.1 are <u>not</u> equivalent! This leads to an interesting family of generalizations of the Fefferman result. But we shall not dwell into these matters here and refer the reader to the original work of Herz.

We should point out however that the functions satisfying Herz's condition $\|\varphi\|_{BMO_p^+} < \infty$ are not necessarily exponentially integrable. Indeed, for $p < 2$ they need not even be square integrable!

III.3 $\mathcal{H}_1$, <u>the   L log L   class and an inequality of Doob</u>.

For any martingale $\{f_\nu\}$ $\nu = 0, 1, \ldots, n$ we have

$$\text{III 3.1} \qquad E\left((f_n^* - 1)^+ \right) \leq E\left(|f_n| \log^+ |f_n| \right) \quad (*)$$

The standard procedure in establishing this inequality is to derive first

---

(*) $\log^+ u = \chi(u > 1) \log u$.

III 3.2 $\qquad \lambda P[f_n^* \geq \lambda] \leq E\left(|f_n| \, \chi(f_n^* \geq \lambda)\right)$,

by a stopping time argument, then dividing III 3.2 by λ and integrating (with respect to λ) from 1 to ∞. (See [11] p. 317-318).

However, we shall see here that III 3.1, III 3.2 as well as II 5.6 and the inequalities in Lemma I.4.2 are all immediate consequences of a general inequality for submartingales which can be stated as follows:

<u>Theorem III 3.1</u> <u>Let</u> $\{P_\nu\}$ $\nu = 1, 2, \ldots, n$ <u>be a sequence of non-negative integrable functions such that</u>

1) $\mathcal{F}(P_\nu) \subset \mathcal{F}_\nu$,

III 3.4

2) $E(P_\nu | \mathcal{F}_{\nu-1}) \geq P_{\nu-1}$,

<u>then, if</u> $m(t)$ <u>is non-negative and non-decreasing in</u> $[0, \infty)$, <u>we have</u>

III 3.5 $\qquad E\left(\int_0^{P_n^*} t \, dm(t)\right) \leq E\left(P_n \, m(P_n^*)\right)$. (*)

(*) As usual $P_\nu^* = \max_{\mu \leq \nu} P_\mu$

III.3

Proof. We can set $P_0 = 0$ without affecting III 3.4. This gives

$$E\left(\int_0^{P_n^*} t\, dm(t)\right) = \sum_{\nu=1}^n E\left(\int_{P_{\nu-1}^*}^{P_\nu^*} t\, dm(t)\right) \leq$$

$$\leq \sum_{\nu=1}^n E\left(P_\nu^*[m(P_\nu^*) - m(P_{\nu-1}^*)]\right).$$

Now, note that $P_\nu^* > P_{\nu-1}^* \Rightarrow P_\nu^* = P_\nu$, thus

$$E\left(\int_0^{P_n^*} t\, dm(t)\right) \leq \sum_{\nu=1}^n E\left(P_\nu [m(P_\nu^*) - m(P_{\nu-1}^*)]\right)$$

Since $m(P_\nu^*) - m(P_{\nu-1}^*)$ is $\mathfrak{F}_\nu$-measurable, III 3.4 ,2) yields

$$E\left(\int_0^{P_n^*} t\, dm(t)\right) \leq \sum_{\nu=1}^n E\left(P_n [m(P_\nu^*) - m(P_{\nu-1}^*)]\right)$$

and this clearly gives III 3.5.

Note. To obtain III 3.2 we set of course $P_\nu = |f_\nu|$ and $m(t) = \log^+ t$. To obtain III 3.2 we set $P_\nu = |f_\nu|$ and $m(t) = \chi(t \geq \lambda)$.

The inequality of lemma I.4.2, is obtained with the same choice of P_ν but using $m(t) = t^{p-1}$ and Hölder's inequality.

It should be pointed out that, III 3.5 itself can also be derived, from Doob's classical inequality III 3.2 upon multiplying by $dm(\lambda)$ and integrating.

Let us now go back to III 3.1. Following an argument of Doob, we note that when $b > a > 0$

$$a \log b = a \log a + b \, \frac{\log b/a}{b/a}$$

$$\leq a \log a + b/e$$

and therefore

$$a \log^+ b \leq a \log^+ a + b/e .$$

Combining with III 3.1 we get then

$$E\left((f_n^*-1)^+\right) \leq E\left(|f_n| \log^+ |f_n|\right) + \frac{1}{e} E(f_n^*).$$

So finally we deduce

$$E\left(f_n^*\right) \leq \frac{e}{e-1} + \frac{e}{e-1} E\left(|f_n| \log^+ |f_n|\right) .$$

Using II 1.1 for $p = 1$ we therefore get

__Theorem III 3.2__ __If__ $f \in L \log L$ __then__ $f \in \mathcal{H}_1$ __and indeed__

III 3.6 $$E\bigl(S(f)\bigr) \le \bigl(2 + \sqrt{5}\bigr)\left(\frac{e}{e-1} + \frac{e}{e-1}\ E\bigl(|f|\ \log^+|f|\bigr)\right)$$

It is thus seen that the work of Burkholder-Gundy and B. Davis give a new light to Doob's classical result.

However, Gundy [18] following up on a work of Stein [27] and Burkholder [4], went further and showed that at least for a certain class of non-negative functions the result in III 3.6 can be reversed!

To present this material and for some later considerations we need to make some definitions. Let us say that a sequence of random variables $\{Q_n\}$ is of "__predictable size__" (with respect to $\{\mathcal{F}_\nu\}$) if and only if there is an adapted non decreasing sequence of random variables $\{\lambda_\nu\}$ such that

III 3.7

a) $|Q_n| \le \lambda_{n-1}, \quad \forall\ n \ge 1$

b) $E(\lambda_\infty) < \infty$.

To be brief, we shall say that a random variable f is "__predictable__" if and only if the sequence $f_n = E(f|\mathcal{F}_n)$ is of predictable size.

Clearly if f is predictable then $f \in \mathcal{H}_1$. The following theorem gives a slight strengthening of a result obtained by Gundy in [18].

__Theorem III 3.3__   When $\mathcal{F}_0 = (\phi, \Omega)$, _every non-negative predictable function in_ $\mathcal{H}_1$ _is also in the_ L log L _class_.

The crucial step in the proof of this result is an inequality for supermartingales which in a sense is the converse of III 3.5, namely:

__Theorem III 3.4__   _If_ $\{Q_\nu\}$ _is a sequence of_ __random variables__ _such that_

III 3.8
 a) $\mathcal{F}(Q_\nu) \subset \mathcal{F}_\nu$,
 b) $E(Q_\nu | \mathcal{F}_{\nu-1}) \leq Q_{\nu-1}$,
 c) $Q_\nu \leq \lambda_{\nu-1}$,

where $\{\lambda_\nu\}$ $\lambda_0 \geq 0$ _is a non-decreasing sequence of_ __random variables__ _adaptable to_ $\{\mathcal{F}_\nu\}$, _then, for any non-decreasing function_ $m(t)$ _we have_

III 3.9 $$E\left(\int_0^{\lambda_n} t\, dm(t) \Big| \mathcal{F}_1\right) \geq E\left(Q_n [m(\lambda_n) - m(\lambda_0)] \Big| \mathcal{F}_1\right)$$

__Proof__. Since $\mathcal{F}(\lambda_\nu) \subset \mathcal{F}_\nu$

$$E\left(Q_n [m(\lambda_n) - m(\lambda_0)] \big| \mathcal{F}_\nu\right) = \sum_{\nu=1}^n E\left(E(Q_n | \mathcal{F}_\nu) [m(\lambda_\nu) - m(\lambda_{\nu-1})] \Big| \mathcal{F}_1\right).$$

Thus, using first III 3.8 b) then III 3.8 c), we get

$$E\left(Q_n\left[m(\lambda_n) - m(\lambda_0)\right]|\mathcal{F}_1\right) \leq \sum_{\nu=1}^{n} E\left(Q_\nu\left[m(\lambda_\nu) - m(\lambda_{\nu-1})\right]|\mathcal{F}_1\right) \leq$$

$$\leq \sum_{\nu=1}^{n} E\left(\lambda_{\nu-1}\left[m(\lambda_\nu) - m(\lambda_{\nu-1})\right]|\mathcal{F}_1\right) \leq$$

$$\leq E\left(\int_0^{\lambda_n} t\, dm(t)\,|\mathcal{F}_1\right) \quad \text{Q.E.D.}$$

To prove Theorem III 3.3 we set $P_n = E(f|\mathcal{F}_n)$ and $m(t) = \log^+ t$ in III 3.9 and get

III 3.10 $\quad E\left(f_n \log^+ \lambda_n |\mathcal{F}_1\right) \leq f_1 \log^+ \lambda_0 + E\left((\lambda_n - 1)^+ |\mathcal{F}_1\right)$

The assumption $\mathcal{F}_0 = (\phi, \Omega)$, upon taking expectations, gives

$$E\left(f_n \log^+ f_n\right) \leq E(f) \log^+ \lambda_0 + E\left((\lambda_n - 1)^+\right) ,$$

and this implies that $f \in L \log L$ when $E(\lambda_\infty) < \infty$.

<u>Remark III 3.1</u> We should observe that the hypothesis "$\mathcal{F}_0 = (\phi, \Omega)$" cannot be omitted altogether in Theorem III 3.3. Indeed, when

$\mathfrak{F}_0 = \mathfrak{F}_1 = \ldots = \mathfrak{F}_n = \ldots = \mathfrak{F}$ then f is clearly predictable and in H_1 if and only if it is in L_1.

Perhaps, the hypothesis "$\mathfrak{F}_0 = (\phi, \Omega)$" is best replaced by "$f_0 \in L \log L$". In fact, if we condition III 3.10 with respect to $\mathfrak{F}_0$ we easily get

$$E\left(f_n \log^+ f_n \mid \mathfrak{F}_0\right) \leq f_0 \log^+ f_0 + \frac{\lambda_0}{e} + E\left((\lambda_n - 1)^+ \mid \mathfrak{F}_0\right)$$

Thus $f \in L \log L$ immediately follows when $E(\lambda_\infty) < \infty$ and $E\left(f_0 \log^+ f_0\right) < \infty$.

Some further remarks are necessary here since Gundy's paper [18], although it contains such nice results, is in places somewhat obscure.

<u>Remark III 3.2</u>　First of all Gundy deals not with predictable sequences but with what he calls "L_∞-regular" martingales.

His definition goes as follows: $\{f_n\}$ is "L_∞-regular" if and only if

$$\Delta f_n = \theta_{n-1} d_n$$

where

III 3.11
a) $\mathcal{F}(\theta_n) \subset \mathcal{F}_n$
b) $E(d_n|\mathcal{F}_{n-1}) = 0$
c) $E(d_n^2|\mathcal{F}_{n-1}) = 1$
d) $|d_n| \leq d$ (= const.) $\forall$ n.

Now, it is quite easy to show that when $f_n \geq 0$ and $f^* \in L_1$ such a sequence is predictable. Indeed, under these hypotheses we can prove

III 3.12 $$f_n \leq f_{n-1}(1 + 2d^2).$$

To see this, note that since $E(\Delta f_n|\mathcal{F}_{n-1}) = 0$ then

$$E\left(|\Delta f_n| \big| \mathcal{F}_{n-1}\right) = 2 E\left([\Delta f_n]^- \big| \mathcal{F}_{n-1}\right).$$

However, the relation

$$[\Delta f_n]^+ - [\Delta f_n]^- = f_n - f_{n-1}$$

gives $[\Delta f_n]^- \leq f_{n-1}$ and we deduce

$$E\left(|\Delta f_n| \big| \mathcal{F}_{n-1}\right) \leq 2 f_{n-1}$$

In our case we have then

$$|\Delta f_n| \leq |\theta_{n-1}| \; d \leq |\theta_{n-1}| \; d \; E(d_n^2|\mathfrak{F}_{n-1}) \leq d^2 \; E(|\Delta f_n||\mathfrak{F}_{n-1}) \leq 2d^2 \; f_{n-1},$$

and III 3.12 necessarily follows.

Curiously enough one of the most important applications of Gundy's result is to martingales relative to "regular" sequences $\{\mathfrak{F}_n\}$. However, for the latter it is easier to obtain III 3.12 directly than showing L_∞-regularity.

By definition $\{\mathfrak{F}_n\}$ "regular" means that each $\mathfrak{F}_n$ is atomic and is obtained from $\mathfrak{F}_{n-1}$ by splitting some of its atoms in a manner that the ratio of the measure of an atom of $\mathfrak{F}_{n-1}$ to any of its parts in $\mathfrak{F}_n$ remains bounded by a fixed constant C independent of n.

This is the case for instance for dyadic partitions of the unit cube in n-dimensional space.

Remark III 3.3   It might be appropriate at this point that we derive the B. Davis [10] decomposition for H_1 functions. This decomposition was the crucial step in Davis' original proof of the inequalities II 1.1 and II 1.2 for $p = 1$.

Here it should give us an interesting view of the significance of the hypothesis that f be predictable.

To this end let us define as G the subset of H_1 consisting of those functions f such that

III 3.11
$$\|f\|_G = \sum_{n=1}^{\infty} E\left(|\Delta f_n|\right) < \infty \; .$$

Let us also introduce as a norm in G precisely the quantity in III 3.11 .

Furthermore, in the subset P of H_1 consisting of predictable functions let us introduce as norm the quantity

$$\|f\|_P = \inf E(\lambda_\infty)$$

where the inf is taken over all non-decreasing adapted sequences $\{\lambda_n\}$ such that $|f_n| \leq \lambda_{n-1} \quad \forall \; n$.

Burgess Davis' result can be stated as follows:

<u>Theorem III 3.5</u> <u>If</u> $f \in H_1$ <u>then</u>

$$f = h + g$$

<u>where</u> $h \in G$, $g \in P$ <u>and</u>

III 3.12
a) $\|h\|_G \leq 8\, E(f^*)$,

b) $\|g\|_P \leq 17\, E(f^*)$.

<u>Proof</u>. Suppose $\lambda_0 \leq \lambda_1 \leq \ldots \leq \lambda_n \ldots$ is an adapted sequence such that

$$|f_n| \leq \lambda_n$$
$$E(\lambda_\infty) < \infty,$$

then we can write

$$\Delta f_n = \Delta f_n\, \chi(\lambda_n > 2\lambda_{n-1}) + \Delta f_n\, \chi(\lambda_n \leq 2\lambda_{n-1})$$

Burgess Davis' idea is to set

$$h = \sum_{\nu=1}^{\infty} \left[\Delta f_\nu\, \chi(\lambda_\nu > 2\lambda_{\nu-1}) - E\left(\Delta f_\nu\, \chi(\lambda_\nu > 2\lambda_{\nu-1}) \mid \mathcal{F}_{\nu-1}\right) \right],$$

$$g = \sum_{\nu=1}^{\infty} \left[\Delta f_\nu\, \chi(\lambda_\nu \leq 2\lambda_{\nu-1}) - E\left(\Delta f_\nu\, \chi(\lambda_\nu \leq 2\lambda_{\nu-1}) \mid \mathcal{F}_{\nu-1}\right) \right].$$

Indeed, when $\lambda_\nu > 2\lambda_{\nu-1}$ then $\lambda_\nu \leq 2(\lambda_\nu - \lambda_{\nu-1})$ and we must have

$$|\Delta f_\nu|\, \chi(\lambda_\nu > 2\lambda_{\nu-1}) \leq 2\lambda_\nu\, \chi(\lambda_\nu > 2\lambda_{\nu-1}) \leq 4\, (\lambda_\nu - \lambda_{\nu-1}).$$

Thus

$$\sum_{\nu=1}^{n} |\Delta h_\nu| \le 4 \lambda_n + 4 \sum_{\nu=1}^{n} E(\lambda_\nu - \lambda_{\nu-1} | \mathcal{F}_{\nu-1}) .$$

This immediately gives

$$\|h\|_G \le 8 \sum_{\nu=1}^{\infty} E(\lambda_\nu - \lambda_{\nu-1}) \le 8 E(\lambda_\infty).$$

On the other hand we have:

$$|\Delta f_\nu \, X(\lambda_\nu \le 2\lambda_{\nu-1})| \le 2\lambda_\nu \, X(\lambda_\nu \le 2\lambda_{\nu-1}) \le 4 \lambda_{\nu-1} .$$

This gives

$$|\Delta g_\nu| \le 8 \lambda_{\nu-1} .$$

So finally we get

$$|g_n| \le |g_{n-1}| + 8 \lambda_{n-1} \le |f_{n-1}| + |h_{n-1}| + 8 \lambda_{n-1} \le$$

$$\le 9 \lambda_{n-1} + 4 \lambda_{n-1} + 4 \sum_{\nu=1}^{n-1} E(\lambda_\nu - \lambda_{\nu-1} | \mathcal{F}_{\nu-1}) .$$

and we obtain

$$\|g\|_{\mathcal{P}} \leq 17\ E(\lambda_\infty).$$

Of course, if $E(f^*) < \infty$ we can take $\lambda_n = f_n^*$ so III 3.12 a) and b) immediately follow.

<u>Remark III 3.4</u>   There are cases in which $\mathcal{H}_1$ and $\mathcal{P}$ coincide. This fact was noticed by Gundy$^{(*)}$ in the case $\{\mathcal{F}_n\}$ is regular.

Indeed, Gundy also states that when f has an expansion of the form

III 3.14 $$f = \sum_{\nu=1}^{\infty} \theta_{\nu-1}\, d_\nu$$

with $\{\theta_\nu\}$ and $\{d_\nu\}$ satisfying III 3.11 then $f \in \mathcal{H}_1$ implies $f \in \mathcal{P}$.

Actually, we can show here that all we need to draw the same implication is the existence of a constant c such that

III 3.15 $$|\Delta f_n| \leq c\ E(|\Delta f_n|\,|\,\mathcal{F}_{n-1}),\quad \forall\ n \geq 1.$$

$^{(*)}$Personal communication.

To see this, note that since $|\Delta f_n| \le 2f_{n-1}^* + f_n^* - f_{n-1}^*$,
III 3.15 implies

$$|\Delta f_n| \le 2c\, f_{n-1}^* + c\, E(f_n^* - f_{n-1}^* | \mathcal{F}_{n-1}).$$

Thus we get

$$|f_n| \le (1 + 2c)\, f_{n-1}^* + c \sum_{\nu=0}^{n-1} E(f_{\nu+1}^* - f_\nu^* | \mathcal{F}_\nu),$$

and we must conclude that

Theorem III 3.6 _If_ f _is in_ $\mathcal{H}_1$ _and satisfies_ III 3.15 _then_ f _is also in_ ρ _and_

$$\|f\|_\rho \le (1 + 3c)\, E(f^*).$$

To derive Gundy's assertions we need only show that when $\{\mathcal{F}_n\}$ is regular or more generally when f has an expansion of the form III 3.14 an inequality of the form III 3.15 necessarily holds.

In both cases this follows almost immediately. In fact, suppose first that we have III 3.14 with III 3.11 then

$$|\Delta f_n| \le d\, |\theta_{n-1}| \le d\, |\theta_{n-1}|\, E\!\left(d_n^2 | \mathfrak{F}_{n-1}\right) \le d^2\, E\!\left(|\Delta f_n|\,|\mathfrak{F}_{n-1}\right),$$

and we have III 3.15 with $c = d^2$.

In case $\{\mathfrak{F}_n\}$ is <u>regular</u>, then by definition there is a constant $c > 0$ such that for any two atoms $A \in \mathfrak{F}_{n-1}$ and $B \in \mathfrak{F}_n$ with $A \supset B$ we have

$$\frac{P(A)}{P(B)} \le c.$$

But then, in B we have

$$|f_n - f_{n-1}| = \frac{1}{P(B)} \int_B |f_n - f_{n-1}| \le \frac{P(A)}{P(B)} \frac{1}{P(A)} \int_A |f_n - f_{n-1}|.$$

But this is III 3.15 again! So we can state

<u>Theorem III 3.7</u>   When $\{\mathfrak{F}_n\}$ is regular

$$\mathcal{H}_1 = \mathcal{P}.$$

III 4   Potentials, and a "dual" to Doobs inequality

Some of the material presented in this section has been inspired by the recent survey of Meyer on "Martingales and stochastic integrals" [24].

We shall be concerned with $\{\mathcal{F}_n\}$-supermartingales. To be specific let $\{Q_n\}$ be a sequence of integrable random variables such that for $\nu = 0, 1, \ldots$

III 4.1
a) $Q_\nu \geq 0$,
b) $\mathcal{F}(Q_\nu) \subset \mathcal{F}_\nu$,
c) $Q_\nu - E(Q_{\nu+1} | \mathcal{F}_\nu) \geq 0$.

It is easy to see that for each $n \geq 1$ we have

$$Q_n = E(Q_1 | \mathcal{F}_0) + \sum_{\nu=1}^{n} \left[Q_\nu - E(Q_\nu | \mathcal{F}_{\nu-1}) \right] - \sum_{\nu=1}^{n-1} E(Q_\nu - Q_{\nu+1} | \mathcal{F}_\nu) .$$

In other words, the following identity holds

III 4.2
$$Q_n = M_n - A_{n-1}$$

where

$$M_n = E(Q_1|\mathcal{F}_0) + \sum_{\nu=1}^{n} \left[Q_\nu - E(Q_\nu|\mathcal{F}_{\nu-1})\right]$$

is a martingale and

$$A_n = \sum_{\nu=1}^{n} E\left(Q_\nu - Q_{\nu+1}|\mathcal{F}_\nu\right) \quad (A_0 = 0)$$

is a non-negative increasing process.

This is the classical decomposition of supermartingales due to Doob. From III 4.2 we deduce that M_n is also non-negative and indeed since

$$E(M_n) = E(Q_1) \leq E(Q_0) \; ,$$

$\{M_n\}$ is L_1-bounded.

From the martingale theorem we deduce that there is an integrable function M_∞ such that

$$\lim_{n\to\infty} M_n = M_\infty \quad \text{a.s.}$$

Furthermore, since A_n is non-decreasing and

III.4

$$E(A_n) = E(Q_1) - E(Q_{n+1}) \le E(Q_1)$$

the function $A_\infty = \sup_n A_n$ is also integrable. Putting all this together with III 4.2 we get that Q_n must also have a limit and indeed

$$Q_\infty = \lim_{n \to \infty} Q_n = M_\infty - A_\infty .$$

This allows us to write III 4.2 in the form

III 4.3 $\quad Q_n = E(Q_\infty | \mathcal{F}_n) + \left[M_n - E(M_\infty | \mathcal{F}_n) \right] + E(A_\infty - A_{n-1} | \mathcal{F}_n) .$

Note now, that since $\{A_n\}$ is non-negative and non-decreasing the random variables A_n are uniformly integrable thus, the M_n's are uniformly integrable if and only if the Q_n's are such. In this case III 4.3 simplifies to

III 4.4 $\quad\quad\quad Q_n = E(Q_\infty | \mathcal{F}_n) + E(A_\infty - A_{n-1} | \mathcal{F}_n) .$

Now, by definition a sequence $\{Q_\nu\}$ satisfying III 4.1 a), b), and c) is said to be a "__potential__" if and only if

$$\lim_{n\to\infty} E(Q_n) = 0 \; .$$

Our considerations then immediately yield the following important result

Theorem III 4.1 A sequence $\{Q_n\}$ <u>satisfying the condition in III 4.1 is a potential if and only if the Q_n's are uniformly integrable and</u>

$$Q_\infty = \lim_{n\to\infty} Q_n = 0 \quad \text{a.s.}$$

<u>Furthermore, if</u> $\{Q_n\}$ <u>is a potential then there is a unique non-negative, non-decreasing adapted process</u> $\{A_n\}$ <u>such that</u> $A_0 = 0$, $A_\infty = \sup_n A_n \in L_1$ <u>and</u>

III 4.5 $$Q_n = E\!\left(A_\infty - A_{n-1} \,\middle|\, \mathfrak{F}_n\right) .$$

Proof. The only thing that remains to be verified is uniqueness, but III 4.5 gives

$$Q_n - E\!\left(Q_{n+1} \,\middle|\, \mathfrak{F}_n\right) = A_n - A_{n-1} \; ,$$

Thus the condition $A_0 = 0$ and III 4.5 determine all successive A_n's.

If $\{Q_n\}$ is a potential the sequence

$$A_n = \sum_{\nu=1}^{n} E(Q_\nu - Q_{\nu+1} | \mathcal{F}_\nu)$$

is usually referred to as the "canonical" increasing process associated to $\{Q_n\}$."

In the last few years a certain number of separate inequalities have been derived yielding estimates on the size of A_∞ in terms of estimates involving the Q_n's, (see in particular [24] Sections 45, 49 and the inequalities 49.2 and 49.3 there).

All these inequalities may be derived from the simple result below, which in a sense is "dual" to Theorem III 3.1, namely;

Theorem III 4.2 Let $\{Q_\nu\}$ be a potential, let $\gamma \geq 0$ be an integrable function such that

III 4.6 $\qquad Q_n \leq E(\gamma | \mathcal{F}_n) \qquad n = 0, 1, 2, \ldots$,

and let $m(t)$ be non-negative and non-decreasing in $[0, \infty]$ then, for the canonical increasing process $\{A_n\}$ we have

III 4.7 $\qquad E\left(\int_0^{A_n} m(t)\, dt \,\Big|\, \mathfrak{F}_1\right) \leq E\left(\gamma\, m(A_n)\,\big|\,\mathfrak{F}_1\right)\; .$

Proof. We start with the trivial observation

$$\int_0^{A_n} m(t)\, dt = \sum_{\nu=1}^n \int_{A_{\nu-1}}^{A_\nu} m(t)\, dt \leq \sum_{\nu=1}^n m(A_\nu)(A_\nu - A_{\nu-1})\; ,$$

thus setting $p_1 = m(A_1)$ and

$$p_\nu = m(A_\nu) - m(A_{\nu-1}) \qquad \nu = 2, 3, \ldots, n$$

we get

$$\int_0^{A_n} m(t)\, dt \leq \sum_{\nu=1}^n \left(\sum_{\mu=1}^\nu p_\mu\right)(A_\nu - A_{\nu-1}) = \sum_{\mu=1}^n p_\mu \sum_{\nu=\mu}^n (A_\nu - A_{\nu-1})\; .$$

Since $\mathfrak{F}(p_\nu) \subset \mathfrak{F}_\nu$ we then get, using III 4.6

III.4

$$E\left(\int_0^{A_n} m(t)\,dt \Big| \mathcal{F}_1\right) \leq \sum_{\mu=1}^n E\left(p_\mu E(A_n - A_{\mu-1}|\mathcal{F}_\mu)\Big|\mathcal{F}_1\right) =$$

$$= \sum_{\mu=1}^n E\left(p_\mu Q_\mu|\mathcal{F}_1\right) \leq \sum_{\mu=1}^n E\left(p_\mu \gamma_\mu|\mathcal{F}_1\right) =$$

$$= E\left(\gamma\, m(A_n)\Big|\mathcal{F}_1\right) \qquad \text{Q.E.D.}$$

Note. The above proof would yield just as well

III 4.8
$$E\left(\int_0^{A_n - A_{\nu-1}} m(t)\,dt \Big| \mathcal{F}_\nu\right) \leq E\left(\gamma\, m(A_n - A_{\nu-1})\Big|\mathcal{F}_\nu\right)$$

$\forall\ \nu \geq 1$.

It is also interesting to note that when $m(0) = 0$ Theorem III 4.2 can be derived from Theorem III 3.4. Indeed, III 4.6 implies

$$R_\nu = E\left(A_n - \gamma|\mathcal{F}_\nu\right) \leq A_{\nu-1} \qquad \nu = 1, 2, \ldots, n\ .$$

Thus, using III 3.9 with R_ν and A_ν replacing Q_ν and λ_ν respectively we get

$$E\left(\int_0^{A_n} t\,dm(t) \Big| \mathcal{F}_1\right) \geq E\left((A_n - \gamma)\, m(A_n)\Big|\mathcal{F}_1\right)$$

and III 4.7 immediately follows after an integration by parts.

We shall now derive some applications. The first one will be an inequality of Burkholder, Davis and Gundy. This can be stated as follows.

Let $\{\varepsilon_\nu\}$ be a sequence of non-negative random variables, let $\Phi(u)$ be convex in $[0, \infty]$ and such that

$$\text{III 4.9} \qquad p = \sup_{u>0} \frac{u\,\Phi'(u)}{\Phi(u)} < \infty ,$$

then there is a constant C_p depending only on p such that

$$\text{III 4.10} \qquad E\!\left(\Phi\!\left(\sum_{\nu=1}^{\infty} E(\varepsilon_\nu | \mathcal{F}_\nu)\right)\right) \leq C_p\, E\!\left(\Phi\!\left(\sum_{\nu=1}^{\infty} \varepsilon_\nu\right)\right).$$

The original proof of III 4.10 (see [5] .) is somewhat difficult and in the intricacies one loses track of C_p. Recently (see [14]) we have shown that III 4.10 can be derived from the inequality of Theorem III 3.1. This yielded III 4.10 with $c_p = p^{2p}$. We shall not repeat the full argument here since we can do even better now. However, it is worthwhile to illustrate the basic idea in the special case $\Phi(u) = u^p$ ($p > 1$). To this end let γ be a non-negative "test" function and note that

$$E\left(\gamma \sum_{\nu=1}^{\infty} E(\epsilon_\nu | \mathcal{F}_\nu)\right) = \sum_{\nu=1}^{\infty} E(\gamma_\nu \epsilon_\nu) \le E\left(\gamma^* \sum_{\nu=1}^{\infty} \epsilon_\nu\right).$$

By Hölder's inequality we then get

$$E\left(\gamma \sum_{\nu=1}^{\infty} E(\epsilon_\nu | \mathcal{F}_\nu)\right) \le \|\gamma^*\|_q \left\|\sum_{\nu=1}^{\infty} \epsilon_\nu\right\|_p$$

where $\frac{1}{p} + \frac{1}{q} = 1$. Thus, by Lemma I 4.2 we obtain

$$E\left(\gamma \sum_{\nu=1}^{\infty} E(\epsilon_\nu | \mathcal{F}_\nu)\right) \le p \|\gamma\|_q \left\|\sum_{\nu=1}^{\infty} \epsilon_\nu\right\|_p.$$

In other words in this case III 4.10 holds with

$$C_p = p^p.$$

In Meyer's notes [24] (see page II 3.1) yet another proof of III 4.10 is given. The basic idea, which is apparently due to Neveu, is to note that for each $n \ge 1$ we have

$$E\left(\sum_{\nu=n}^{\infty} \epsilon_\nu | \mathcal{F}_n\right) = E\left(\sum_{\nu=n}^{\infty} E(\epsilon_\nu | \mathcal{F}_\nu) | \mathcal{F}_n\right).$$

In other words, if we set

III 4.11
$$A_n = \sum_{\nu=1}^{n} E(\epsilon_\nu | \mathcal{F}_\nu),$$

$$B_n = \sum_{\nu=1}^{n} \epsilon_\nu$$

the sequence $\{A_n\}$ is none other than the canonical process associated to the potential

III 4.12 $\quad Q_n = E\left(B_\infty - B_{n-1} | \mathcal{F}_n\right) = E\left(A_\infty - A_{n-1} | \mathcal{F}_n\right).$

If we combine Neveu's observation with our inequality III 4.7 we obtain the following remarkably sharp inequality:

<u>Theorem III 4.3</u> <u>Let</u> $\{\epsilon_\nu\}$ <u>be a sequence of non-negative random variables and let</u> $\Phi(u)$ <u>be convex in</u> $[0, \infty)$ <u>and such that</u>

III 4.13 $\quad p = \sup_{u>0} \dfrac{u\,\Phi'(u)}{\Phi(u)} < \infty$

<u>then</u>

III 4.14 $\quad E\left(\Phi\left(\sum_{\nu=1}^{\infty} E(\epsilon_\nu|\mathcal{F}_\nu)\right)\right) \leq p^{p+1} E\left(\Phi\left(\sum_{\nu=1}^{\infty} \epsilon_\nu\right)\right)$.

Proof. By definition $p \geq 1$ and since the case $p = 1$ is trivial we shall assume $p > 1$. This given, for convenience we write

$$\Phi(u) = \int_0^u \varphi(t)\, dt \ .$$

We also let

$$\Psi(v) = \int_0^v \psi(s)\, ds$$

where $\psi(s)$ is the inverse of $v = \varphi(u)$. In otherwords, $\Psi(v)$ is the "conjugate" of $\Phi(u)$ in the sense of Young. (See [22]).

Now, it is not difficult to show that III 4.13 implies

III 4.15
a) $\Phi(pu) \leq p^p \Phi(u)$
b) $\Psi\left(\dfrac{v}{p}\right) \leq \dfrac{1}{p} \Psi(v)$
c) $\Psi\big(\varphi(u)\big) \leq (p-1)\, \Phi(u)$.

Note that we have

$$Q_n = E\left(A_\infty - A_{n-1} \mid \mathcal{F}_n\right) = E\left(B_\infty - B_{n-1} \mid \mathcal{F}_n\right) \leq E\left(B_\infty \mid \mathcal{F}_n\right)$$

thus III 4.7 with $\gamma = B_\infty$ and $m(u) = \varphi(u)$ immediately gives

III 4.16 $$E\left(\Phi(A_\infty)\right) \leq E\left(B_\infty \, \varphi(A_\infty)\right)$$

So by Young's inequality, written in the form

$$u\,v \leq \Phi(pu) + \Psi\left(\frac{v}{p}\right)$$

we obtain

$$E\left(\Phi(A_\infty)\right) \leq E\left(\Phi(p\,B_\infty)\right) + E\left(\Psi\left(\frac{\varphi(A_\infty)}{p}\right)\right)$$

and the inequalities in III 4.15 give

$$E\left(\Phi(A_\infty)\right) \leq p^p \, E\left(\Phi(B_\infty)\right) + \frac{1}{p}\,(p-1)\,E\left(\Phi(A_\infty)\right)$$

In other words

$$\frac{1}{p} E\big(\Phi(A_\infty)\big) \le p^p\, E\big(\Phi(B_\infty)\big)$$

and this combined with III 4.11 yields III 4.14 as desired.

<u>Remark III 4.1</u> In the particular case that the potential Q_n has the form

III 4.17 $\qquad Q_n = E(B_\infty - B_{n-1} | \mathcal{F}_n)$

with $B_n \uparrow B_\infty$, the inequality III 4.16 can be derived more quickly than III 4.7.

Indeed, since $\mathcal{F}(A_\nu) \subset \mathcal{F}_\nu$ and $A_\nu - A_{\nu-1} = E(B_\nu - B_{\nu-1} | \mathcal{F}_\nu)$ we have

$$E\left(\int_0^{A_n} \varphi(u)\, du\right) \le \sum_{\nu=1}^n E\Big(\varphi(A_\nu)\, \big(A_\nu - A_{\nu-1}\big)\Big) = \sum_{\nu=1}^n E\Big(\varphi(A_\nu)\, \big(B_\nu - B_{\nu-1}\big)\Big) \le$$
$$\le E\Big(\varphi(A_n)\, B_n\Big).$$

<u>Remark III 4.2</u> Note that if the potential

$$Q_n = E\Big(A_\infty - A_{n-1} \big| \mathcal{F}_n\Big)$$

is bounded say by a constant B, then III 4.7 with $m(t) = t^{p-1}$ and $\gamma = B$ gives

$$E\left(A_n^p\right) \le p\, E\left(A_n^{p-1}\right).$$

Also for $m(t) = e^t$ we easily obtain

$$E\left(e^{A_n}\right) \le \frac{1}{1-B}.$$

These are the inequalities in Section 45 of [24]. But, clearly in this case the sequence $\{A_n\}$ is BMO(B) so these estimates also follow from our previous considerations.

It is interesting to note however, that the function A_∞ itself is BMO!

To see this note that since $A_n \uparrow A_\infty$

$$A_{n-1} - E\left(A_\infty | \mathfrak{F}_{n-1}\right) \le A_\infty - E\left(A_\infty | \mathfrak{F}_{n-1}\right) \le A_\infty - A_{n-1},$$

thus

$$\left| A_\infty - E\left(A_\infty | \mathfrak{F}_{n-1}\right) \right| \le A_\infty - A_{n-1} + E\left(A_\infty - A_{n-1} | \mathfrak{F}_{n-1}\right),$$

which gives

$$\|A_\infty\|_{BMO_1} \leq 2B.$$

The simplest way to obtain a bounded potential is to take a sequence $B_n \uparrow B_\infty$ such that

$$\|B_\infty\|_\infty < \infty$$

and set

$$Q_n = E\left(B_\infty - B_{n-1} \big| \mathfrak{F}_n\right).$$

In analogy with Theorem III 4.1 one would conjecture that every bounded potential can be represented in this form. However, to this date we have no proof of this fact.

<u>Remark III 4.3</u> We should point out that the inequality of Burkholder, Davis and Gundy (Theorem III 4.3) reverses direction when we take <u>concave</u> functions rather than <u>convex</u> ones.

We shall carry this out only when

$$\varphi(u) = u^{1/p} \quad (p > 1)$$

since the general case is not needed here. Let then

$$\Phi(u) = \int_0^u \frac{dt}{\varphi(t)}$$

where $\varphi(t)$ increases in $[0, \infty]$. Set as before

III 4.18
$$B_n = \sum_{\nu=1}^n \varepsilon_\nu \; ,$$

$$A_n = \sum_{\nu=1}^n E(\varepsilon_\nu | \mathfrak{F}_\nu)' \; .$$

We then have

$$E\left(\int_0^{A_n} \frac{dt}{\varphi(t)}\right) \geq \sum_{\nu=1}^n E\left(\frac{1}{\varphi(A_\nu)}(A_\nu - A_{\nu-1})\right) = \sum_{\nu=1}^n E\left(\frac{1}{\varphi(A_\nu)}(B_\nu - B_{\nu-1})\right) \geq$$

$$\geq E\left(\frac{1}{\varphi(A_n)} B_n\right) \; .$$

Setting

$$\varphi(u) = p \, u^{1-1/p} = p \, u^{1/q} \quad \left(\frac{1}{p} + \frac{1}{q} = 1\right)$$

we get

$$\frac{1}{p} \, E\left(\frac{B_n}{A_n^{1/q}}\right) \leq E\left(A_n^{1/p}\right).$$

A use of Hölder's inequality then gives

$$E\left(B_n^{1/p}\right) \leq \left[E\left(\frac{B_n}{A_n^{1/q}}\right)\right]^{1/p} \left[E\left(A_n^{1/p}\right)\right]^{1/q} \leq$$

$$\leq \left[p \, E\left(A_n^{1/p}\right)\right]^{1/p} \left[E\left(A_n^{1/p}\right)\right]^{1/q}.$$

So we can state

<u>Theorem III 4.4</u>   <u>If</u> $\{\epsilon_\nu\}$ <u>is a sequence of non-negative random variables
then for</u> $0 < p \leq 1$ <u>we have</u>

III 4.19 $\qquad p^p \, E\left(\left[\sum_{\nu=1}^{\infty} \epsilon_\nu\right]^p\right) \leq E\left(\left[\sum_{\nu=1}^{\infty} E(\epsilon_\nu | \mathcal{F}_\nu)\right]^p\right)$

<u>and for</u> $p \geq 1$

III 4.20 $$E\left(\left[\sum_{\nu=1}^{\infty} E(\epsilon_\nu|\mathcal{F}_\nu)\right]^p\right) \leq p^p \, E\left(\left[\sum_{\nu=1}^{\infty} \epsilon_\nu\right]^p\right).$$

III 5   On the possibility of extending the definition of $\mathcal{K}_q$ spaces.

In Chapter II we have essentially proved the following result

Theorem III 5.1   *If* $f \in L_2$ *is such that*

III 5.1 $$E\left(|f - f_{n-1}|^2 \,|\mathcal{F}_n\right) \leq E\left(\gamma^2|\mathcal{F}_n\right) \quad (\forall \; n \geq 1)$$

with $\gamma \in L_q$ $(q \geq 2)$ *then* $f \in L_q$ *and*

III 5.2 $$\|f\|_q \leq 5 \, q \, \sqrt{2 \, \frac{q}{q-1}} \, \|\gamma\|_q .$$

Indeed, the inequality

III 5.3 $$|E(\varphi_n \, f_n)| \leq \sqrt{\frac{2}{p}} \, \|S_n(\varphi)\|_p \, \|f\|_{\mathcal{K}_q}$$

proved in Section I.3 combined with

III 5.4 $\qquad \|S_n(\varphi)\|_p \le 5 \quad \|\varphi_n^*\|_p \le 5q \quad \|\varphi\|_p \qquad (*)$

gives

$$|E(\varphi_n\, f_n)| \le \sqrt{\tfrac{2}{p}}\; 5q \; \|\varphi\|_p \; \|f\|_{\mathcal{H}_q}$$

and this clearly yields II 5.2 .

It stands to reason, in view of the beauty of the inequality in III 5.3 that we should attempt to extend the $\mathcal{H}_q$-norm and $\mathcal{H}_q$ spaces for values of $q \le 2$. To this date, we are aware of no completely satisfactory solution to this problem.

We can, however, extend Theorem III 5.1 without III 5.3 . This we shall do here. In the next chapter we shall also present some sort of extension of III 5.3 .

Our result here can be stated as follows:

<u>Theorem III 5.2</u> <u>Let</u> $f \in L_1$ <u>be such that</u> $f_0 = 0$ <u>and</u>

(*)
    This follows from II 2.15 and Lemma I.4.2 .

III 5.5 $\quad E\left(|f - f_{n-1}|\,\big|\,\mathcal{F}_n\right) \leq E\left(\gamma\,\big|\,\mathcal{F}_n\right) \quad \forall\; n \geq 1$

<u>for some</u> $\gamma \in L_p$ $(p>1)$ <u>then</u> f <u>is in</u> L_p <u>and</u>

III 5.6 $\quad \|f^*\|_p \leq 3\; qpe\; \|\gamma\|_p\;.$

This inequality follows immediately from the following lemma which in a sense extends Theorem III 2.1 .

<u>Lemma III 5.1</u> <u>Let</u> $\{A_n\}$ <u>be a sequence of integrable random variables such that</u>

III 5.7
a) $\mathcal{F}(A_n) \subset \mathcal{F}_n$,

b) $E\left(|A_n - A_{\nu-1}|\,\big|\,\mathcal{F}_\nu\right) \leq E\left(\gamma\,\big|\,\mathcal{F}_\nu\right) \quad \forall\; n \geq \nu \geq 1$,

<u>then setting</u> $A_n^* = \max\limits_{1 \leq \nu \leq n} |A_\nu - A_0|$ <u>we have</u>

III 5.8 $\quad E\left(A_n^{*\,p}\,\big|\,\mathcal{F}_1\right) \leq \left(3\; pqe\right)^p E\left(\gamma^p\,\big|\,\mathcal{F}_1\right)$

Proof. We shall follow closely the arguments used in the proof of Theorem III 2.1 . Since there is no loss we assume $A_0 = 0$ and set as in III 2.1

III 5.9
$$A_n^\alpha = \sum_{\mu=1}^{n} \chi(A_{\mu-1}^* \le \alpha;\ A_\mu^* > \alpha)\ A_\mu\ ,$$

let $\beta > \alpha$ and A_n^β be defined in the same way. It is easy to see that we always have

$$\beta\ \chi\!\left(A_n^* > \beta\right) \le |A_n^\beta| \le |A_n^\beta - A_n^\alpha| + |A_n^\alpha|\ \chi\!\left(A_n^* > \beta\right)\ .$$

Indeed, from the definition $A_n^\beta \ne 0$ if and only if $A_n^* > \beta$ and then $|A_n^\beta| > \beta$.

Taking conditional expectations we get

$$\beta\ E\!\left(\chi(A_n^* > \beta)\,|\mathcal{F}_1\right) \le E\!\left(|A_n - A_n^\beta|\,|\mathcal{F}_1\right) + E\!\left(|A_n - A_n^\alpha|\,|\mathcal{F}_1\right) +$$
$$+ E\!\left(|A_n^\alpha|\ \chi(A_n^* > \beta)\,|\mathcal{F}_1\right)$$

Observe then that III 5.7 gives

$$E\left(|A_n - A_n^\beta|\,|\mathcal{F}_1\right) \le \sum_{\mu=1}^{n} E\left(\chi(A_{\mu-1}^* \le \beta;\ A_\mu^* > \beta)\ |A_n - A_\mu|\,|\mathcal{F}_1\right) =$$

$$= \sum_{\mu=1}^{n} E\left(\chi(A_{\mu-1}^* \le \beta;\ A_\mu^* > \beta)\ E\left(|A_n - A_\mu|\,|\mathcal{F}_\mu\right)|\mathcal{F}_1\right)$$

$$\le \sum_{\mu=1}^{n} E\left(\chi(A_{\mu-1}^* \le \beta;\ A_\mu^* > \beta)\ \gamma_\mu|\mathcal{F}_1\right) =$$

$$= E\left(\chi(A_n^* > \beta)\ \gamma|\mathcal{F}_1\right)$$

In a completely analogous way we obtain

$$E\left(|A_n - A_n^\alpha|\,|\mathcal{F}_1\right) \le E\left(\chi(A_n^* > \alpha)\ \gamma|\mathcal{F}_1\right)$$

Finally, we have

$$E\left(|A_n^\alpha|\ \chi(A_n^* > \beta)\,|\mathcal{F}_1\right) = \sum_{\mu=1}^{n} E\left(\chi(A_{\mu-1}^* \le \alpha;\ A_\mu^* > \alpha)\ |A_\mu|\ \chi(A_n^* > \beta)\,|\mathcal{F}_1\right) \le$$

$$\le \sum_{\mu=1}^{n} E\left(\chi(A_{\mu-1}^* \le \alpha;\ A_\mu^* > \alpha)\ |A_{\mu-1}|\ \chi(A_n^* > \beta)\,|\mathcal{F}_1\right) +$$

$$+ \sum_{\mu=1}^{n} E\left(\chi(A_{\mu-1}^* \le \alpha;\ A_\mu^* > \alpha)\ |A_\mu - A_{\mu-1}|\,|\mathcal{F}_1\right)$$

Now, III 5.7 gives $|A_\mu - A_{\mu-1}| = E\left(|A_\mu - A_{\mu-1}|\,\Big|\mathcal{F}_\mu\right) \le \gamma_\mu$ thus

$$E\left(|A_n|^\alpha \chi(A_n^* > \beta) | \mathcal{F}_1\right) \le \alpha \; E\left(\chi(A_n^* > \beta) | \mathcal{F}_1\right) + E\left(\chi(A_n^* > \alpha) \gamma | \mathcal{F}_1\right)$$

Combining these estimates we easily get

$$(\beta - \alpha) \; E\left(\chi(A_n^* > \beta) | \mathcal{F}_1\right) \le 2 \; E\left(\chi(A_n^* > \alpha) \gamma | \mathcal{F}_1\right) + E\left(\chi(A_n^* > \beta) \gamma | \mathcal{F}_1\right)$$

Without trying to be optimal with our constants we set $\beta = \frac{p+1}{p}$ and multiply by α^{p-2} to get

$$\frac{\alpha^{p-1}}{p} \; E\left(\chi\left(\frac{p}{p+1} A_n^* > \alpha\right) | \mathcal{F}_1\right) \le 3\alpha^{p-2} \; E\left(\chi(A_n^* > \alpha) \gamma | \mathcal{F}_1\right) ,$$

an integration with respect to α from 0 to ∞ gives

$$\frac{1}{p^2} \left(\frac{p}{p+1}\right)^p \; E\left(A_n^{*p} | \mathcal{F}_1\right) \le \frac{3}{p-1} \; E\left(A_n^{*p-1} \gamma | \mathcal{F}_1\right) ,$$

or better yet

$$E\left(A_n^{*p} | \mathcal{F}_1\right) \le 3pq \; e \; E\left(A_n^{*p-1} \gamma | \mathcal{F}_1\right) .$$

Now, because of the estimate $|A_\nu - A_{\nu-1}| \le \gamma_\nu$ the integrability of A_n^{*p} is not in question when $\gamma \in L_p$, thus Hölder's inequality yields

$$E(A_n^{*p}|\mathcal{F}_1) \leq 3 \text{ pqe} \left[E(A_n^{*p}|\mathcal{F}_1)\right]^{\frac{1}{q}} \left[E(\gamma^p|\mathcal{F}_1)\right]^{\frac{1}{p}}$$

which implies III 5.8 .

<u>Remark III 5.1</u> It is good to compare our Theorem III 5.2 with a recent result of Fefferman and Stein [13] and with Lemma 3 of John-Nirenberg's paper [21].

Our result here is essentially "<u>in between</u>". Of course, in both these papers the authors are concerned with integrals of the form

$$\frac{1}{|I|} \int_I |f - f_I| dx , \quad f_I = \frac{1}{|I|} \int_I f \, dx$$

where I is a d-dimensional cube. To translate their results in the present setup we are to replace the above expressions by

$$E\left(|f - f_n| \big| \mathcal{F}_n\right) \quad \text{and} \quad E\left(f | \mathcal{F}_n\right)$$

respectively.

If we do this we see that the John-Nirenberg hypothesis rather than III 5.5 is essentially

III 5.9 $$E\left(\left[E\left(|f - f_{n(\omega)}| \big| \mathcal{F}_{n(\omega)}\right)\right]^p\right) \leq C < \infty$$

where C is a fixed constant and $n(\omega)$ is an arbitrary stopping time, i.e. a random variable such that

$$\{\omega: n(\omega) = k\} \in \mathfrak{F}_k \ .$$

It is easy to see that III 5.5 implies III 5.9 and accordingly John & Nirenberg obtain a weaker result namely that f is in weak-L_p. To be precise they derive

$$P\left[|f - f_n| > \lambda \,|\, \mathfrak{F}_n\right] \leq \frac{c}{\lambda^p} \ .$$

where c depends only on C, p and the dimension of the space.

Incidentally, if we use the implications of Lemma III 5.1 in full we can derive just as well as III 5.6 the stronger inequality

III 5.10 $\qquad E\left(\sup_{n \geq \nu} |f_n - f_{\nu-1}|^p \,\Big|\, \mathfrak{F}_\nu\right) \leq (3 \, pqe)^p \, E\left(\gamma^p \,|\, \mathfrak{F}_\nu\right) \ .$

The hypothesis of Fefferman and Stein is considerably stronger than ours. Indeed, they work with what they call the "sharp" function, i.e.

$$f^{\#}(x) = \sup_{I \ni x} \frac{1}{|I|} \int_I |f - f_I| \, dx \ .$$

where the "sup" is taken over cubes centered at x.

They require $f^{\#}$ to be in L_p. This would correspond here to replacing III 5.5 by

$$f^{\#} = \sup_n E\left(|f - f_{n-1}|\,|\mathfrak{F}_n\right) \in L_p \; .$$

Their conclusion is, however, the same as ours, namely that f must be in L_p.

Remark III 5.2   Since the condition

III 5.10 $$E\left(|f - f_{n-1}|\,|\mathfrak{F}_n\right) \le E(\gamma|\mathfrak{F}_n)$$

with $\gamma \in L_\infty$ is equivalent to $f \in BMO$ and with $\gamma \in L_p$ ($p>1$) is equivalent to $f \in \mathcal{H}_p$ we might perhaps guess that III 5.10 with $\gamma \in L_1$ characterizes $\mathcal{H}_1$.

However, this is false. Indeed, we can even exhibit a non-negative f in L_1 such that $|f_n - f_{n-1}| \le 1 \;\; \forall\, n$ which is not in $\mathcal{H}_1$.

Clearly, for such an f we have III 5.10 for

$$E\left(|f - f_{n-1}|\,|\mathfrak{F}_n\right) \le E\left(2f + 1\,|\mathfrak{F}_n\right) \; .$$

This given we might then hope that perhaps the Fefferman-Stein type condition

$$f^{\#} = \sup_n E\left(|f - f_{n-1}|\,\big|\,\mathcal{F}_n\right) \in L_1$$

might characterize $\mathcal{H}_1$. But this also is false in general. Indeed, if $f^{\#} \in L_1$ implies $f \in \mathcal{H}_1$ then since

$$E\left(|f|\,\big|\,\mathcal{F}_n\right) \le E\left(|f - f_{n-1}|\,\big|\,\mathcal{F}_n\right) + |f_{n-1}|$$
$$\le f^{\#} + f^{*}$$

From the Burgess Davis inequality (II 1.2 p = 1) we must then conclude that $f^{\#} \in \mathcal{H}_1$ also implies $|f| \in \mathcal{H}_1$.

But in the case the sequence $\{\mathcal{F}_n\}$ is regular this would force f in the L log L class!

But, we know that in this case there are f's in $\mathcal{H}_1$ which are not in L log L. Thus, for such f's the sharp function cannot be in L_1!

These considerations reveal that whether or not the behavior of the constant in III 5.6 as $p \to 1$ is optimal, it is perfectly consistent with the facts that it should tend to infinity as $p \to 1$.

IV. MARTINGALE TRANSFORM TECHNIQUES

IV. 1 The space $\mathcal{H}_1^-$ and its relation to $\mathcal{P}$ and $\mathcal{H}_1$.

In the work of Burkholder and Gundy a certain number of the results concerning the square function

$$S(f) = \sqrt{\sum_{\nu=1}^{\infty} [\Delta f_\nu]^2}$$

are also obtained for what they call the "conditioned" square function namely

IV 1.1 $$\sigma(f) = \sqrt{\sum_{\nu=1}^{\infty} E\left([\Delta f_\nu]^2 \mid \mathcal{F}_{\nu-1}\right)} \;,$$

Both square functions $S(f)$ and $\sigma(f)$ arise naturally when we write down the Doob representation for the potentials

$$E\left(|f - f_{n-1}|^2 \mid \mathcal{F}_n\right) = E\left(\sum_{\nu=n}^{\infty} [\Delta f]^2 \mid \mathcal{F}_n\right)$$

and

$$E\left(|f - f_n|^2 \mid \mathcal{F}_n\right) = E\left(\sum_{\nu=n+1}^{\infty} [\Delta f_\nu]^2 \mid \mathcal{F}_n\right) \;,$$

for a given f in L_2.

Indeed, if we set

$$S_n(f) = \sqrt{\sum_{\nu=1}^{n} [\Delta f_\nu]^2}$$

$$\sigma_n(f) = \sqrt{\sum_{\nu=1}^{n} E\left([\Delta f_\nu]^2 | \mathfrak{F}_{\nu-1}\right)} \;,$$

it is easily seen that

IV 1.2 $\qquad E\left(|f - f_{n-1}|^2 | \mathfrak{F}_n\right) = E\left(S^2(f) - S_{n-1}^2(f) | \mathfrak{F}_n\right) \;,$

IV 1.3 $\qquad E\left(|f - f_n|^2 | \mathfrak{F}_n\right) = E\left(\sigma^2(f) - \sigma_n^2(f) | \mathfrak{F}_n\right)$

In other words the sequences $\{S_{n-1}(f)\}$ and $\{\sigma_n(f)\}$ are none other than the canonical processes associated to $E\left(|f - f_{n-1}|^2 | \mathfrak{F}_n\right)$ and $E\left(|f - f_n|^2 | \mathfrak{F}_n\right)$ respectively.

The remarkable fact is that $\sigma(f)$ may exist even when f is not square integrable. However, the integrability of $\sigma(f)$ is in general more stringent than the integrability of $S(f)$.

It is clear, in view of the fact that $\sigma(f)$ arises in such a natural way, that parallel to the $\mathcal{H}_p$ spaces we should also consider the spaces

$$\mathcal{H}_p^- = \{f: \sigma(f) \in L_p\}$$

with norms

$$\|f\|_{\mathcal{H}_p^-} = \|\sigma(f)\|_p \ .$$

The last theorem proved in Chapter III provides us immediately with the following important result

<u>Theorem IV 1.1</u>   For $1 \leq p \leq 2$ we have

$$\mathcal{H}_p^- \subset \mathcal{H}_p$$

and for $2 \leq p < \infty$

$$\mathcal{H}_p \subset \mathcal{H}_p^- \ .$$

<u>Indeed, the following inequalities hold</u>

IV 1.4 $\qquad \|S(f)\|_p \leq \sqrt{\dfrac{2}{p}} \ \|\sigma(f)\|_p \qquad 0 \leq p \leq 2 \ ,$

IV 1.5 $$\|\sigma(f)\|_p \leq \sqrt{\frac{p}{2}} \, \|S(f)\|_p \qquad 2 \leq p < \infty .$$

Proof. Just replace p by $p/2$, ε_ν by $[\Delta f_\nu]^2$ and $\mathcal{F}_\nu$ by $\mathcal{F}_{\nu-1}$ in III 4.19 and III 4.20.

In a sense, $\sigma(f)$ is even more "natural" than $S(f)$ itself. Indeed, as we shall see in this chapter, some of the hard fought inequalities obtained for $S(f)$ can be proved for $\sigma(f)$ with the greatest of ease. This is partly due to the fact that $\sigma_n(f)$ is $\mathcal{F}_{n-1}$-measurable, and as a result certain "martingale transform" techniques are available to deal with $\sigma(f)$ which are completely powerless when $S(f)$ is concerned.

In this section we shall introduce these techniques by showing the following result

<u>Theorem IV 1.2</u> <u>Let</u> $f \in L_1$ <u>be such that</u> $f_0 = 0$ <u>and</u>

IV 1.5 $$|f_n| = |E(f|\mathcal{F}_n)| \leq \lambda_{n-1}$$

<u>where</u> $\{\lambda_\nu\}$ <u>is a non-decreasing process adapted to</u> $\{\mathcal{F}_n\}$. <u>Then if</u> $\lambda_\infty = \sup \lambda_n$ <u>is in</u> L_1, f <u>is in</u> $\mathcal{H}_1^-$ <u>and</u>

IV 1.6 $$\|\sigma(f)\|_1 \leq 2 \, E(\lambda_\infty) .$$

Proof. Set

$$g_n = \sum_{\nu=1}^{n} \frac{\Delta f_\nu}{\sqrt{\lambda_{\nu-1}}} .$$

We have

$$g_n = \frac{f_n}{\sqrt{\lambda_{n-1}}} + \sum_{\nu=1}^{n-1} \frac{f_\nu}{\sqrt{\lambda_{\nu-1}}} - \sum_{\nu=1}^{n-1} \frac{f_\nu}{\sqrt{\lambda_\nu}} = \frac{f_n}{\sqrt{\lambda_{n-1}}} + \sum_{\nu=1}^{n-1} \frac{f_\nu}{\sqrt{\lambda_{\nu-1}}} \frac{\sqrt{\lambda_\nu} - \sqrt{\lambda_{\nu-1}}}{\sqrt{\lambda_\nu}}$$

Thus, using IV 1.5

$$|g_n| \leq \sqrt{\lambda_{n-1}} + \sum_{\nu=1}^{n-1} \left(\sqrt{\lambda_\nu} - \sqrt{\lambda_{\nu-1}} \right) \leq 2 \sqrt{\lambda_{n-1}} ,$$

and this implies

IV 1.7 $$E\left(g_n^{*2}\right) \leq 4 \; E(\lambda_{n-1}) .$$

In other words g_n is an L_2-bounded martingale, therefore the limit

$$g = \sum_{\nu=1}^{\infty} \frac{\Delta f_\nu}{\sqrt{\lambda_{\nu-1}}},$$

exists and is in L_2.

But, we have

$$f_n = \sum_{\nu=1}^{n} \Delta g_\nu \; \lambda_{\nu-1}$$

thus

$$\sigma_n^2(f) = \sum_{\nu=1}^{n} E([\Delta g_\nu]^2 | \mathcal{F}_{\nu-1}) \; \lambda_{\nu-1} \leq \lambda_{n-1} \; \sigma_n^2(g)$$

And we obtain

$$E\Big(\sigma_n(f)\Big) \leq E\Big(\sqrt{\lambda_{n-1}} \; \sigma_n(g)\Big) \leq$$

$$\leq \sqrt{E\big(\lambda_{n-1}\big)} \; \sqrt{E\big(\sigma_n^2(g)\big)} =$$

$$= \sqrt{E\big(\lambda_{n-1}\big)} \; \sqrt{E(g_n^2)} \; .$$

Thus using IV 1.7 we finally get

$$E\Big(\sigma_n(f)\Big) \leq 2 \; E(\lambda_{n-1}) \; ,$$

which yields IV 1.6 as $n \to \infty$.

Combining Theorems IV 1.1 and IV 1.2 we can state the following remarkable result

Theorem IV 1.3 For a general sequence of σ-fields $\{\mathcal{F}_n\}$ we have

$$\mathcal{P} \subset \mathcal{H}_1^- \subset \mathcal{H}_1 .$$

and indeed

$$\frac{1}{\sqrt{2}} \, \|f\|_{\mathcal{H}_1} \leq \|f\|_{\mathcal{H}_1^-} \leq 2 \, \|f\|_{\mathcal{P}} .$$

IV 2 The spaces $\mathcal{P}_p$, $\mathcal{H}_p^+$ and a further generalization of Fefferman's inequality.

One of the generalizations of Fefferman's theorem obtained by Carl Herz in [19] is that the dual of $\mathcal{P}$ is none other than the space

$$BMO_1^+ = \{\varphi : \sup_n \|E(|\varphi - \varphi_n| \,|\, \mathcal{F}_n)\|_\infty < \infty \} .$$

with norm

$$\|\varphi\|_{BMO_1^+} = \sup_n \|E(|\varphi - \varphi_n| \,|\, \mathcal{F}_n)\|_\infty .$$

In establishing this fact Herz derives the following beautiful inequality

IV 2.1
$$|E(f\varphi)| \leq 12 \, \|f\|_p \, \|\varphi\|_{BMO_1^+} .$$

Of course, as was the case for Fefferman's inequality, the functional $E(f\varphi)$ need not make sense as a Lebesgue integral. Thus part of the proof of IV 2.1 consists in showing that there is a natural way of extending the definition of $E(f\varphi)$ to all $f \in P$ and $\varphi \in BMO^+$.

In analogy with what we did in I 3, in this section we shall show that an "L_p" version of IV 2.1 holds just as well. Before we can state this result we need to introduce two spaces which will play here the role of H_p and K_q in Section I 3.

To this end let us say that a random variable f is L_p-predictable if and only if there is a sequence $\{\lambda_n\}$ such that

IV 2.2
a) $|f_n| \leq \lambda_{n-1}$
b) $\lambda_0 \leq \lambda_1 \leq \ldots \leq \lambda_n \leq \ldots \leq \lambda_\infty$,
c) $\mathcal{F}(\lambda_\nu) \subset \mathcal{F}_\nu$,
d) $E(\lambda_\infty^p) < \infty$.

We then let ρ_p be the class of L_p-predictable functions f with $f_0 = 0$ and take as norm the functional

$$\|f\|_{\rho_p} = \inf_{\{\lambda_n\}} [E(\lambda_\infty^p)]^{1/p}$$

where the "inf" is taken over all sequences $\{\lambda_n\}$ satisfying IV 2.2.

Emulating our definition of $\mathcal{K}_q$ spaces, for a given $\varphi \in L_1$ we set here

$$\Gamma_\varphi^+ = \left\{ \gamma \in L_1 : E(|\varphi - \varphi_n| | \mathcal{F}_n) \leq E(\gamma | \mathcal{F}_n) \right\} \quad \forall \ n \geq 0,$$

and let

$$\mathcal{K}_q^+ = \left\{ \varphi \in L_1 : \varphi_0 = 0 \ \& \ \inf_{\gamma \in \Gamma_\varphi} [E(\gamma^q)]^{1/q} < \infty \right\}$$

with norm

$$\|\varphi\|_{\mathcal{K}_q^+} = \inf_{\gamma \in \Gamma_\varphi} [E(\gamma^q)]^{1/q}.$$

As we might guess, the space $\mathcal{K}_q^+$ is (for $q = \frac{p}{p-1}$) the dual of

ρ_p. We shall not show this result here and only prove an inequality which contains IV 2.1.

Our result can be stated as follows

<u>Theorem IV 2.1</u> <u>Let</u> φ <u>be in</u> κ_q^+ <u>and</u> f <u>be bounded, then if</u> $Y \in \Gamma_\varphi^+$ <u>and</u> $\{\lambda_n\}$ <u>satisfies IV 2.2 we have</u>

IV 2.3 $\qquad |E(f\varphi)| \leq (4 + 4 \log 2) \, E(\lambda_\infty Y).$

Proof. The argument, which is inspired by C. Herz's proof of IV 2.1, will give us another illustration of the power of the martingale transform techniques.

Since there is no loss we can assume that λ_0 is bounded away from zero, (*) and set for each $n \geq 1$

$$h_n = \sum_{\nu=1}^n \frac{\Delta f_\nu}{\lambda_{\nu-1}}.$$

Now, we see that for $m \leq M$

(*)If worst comes to worst, we can always add an ϵ to each λ_n then at the end let $\epsilon \to 0$.

$$h_M - h_m = \sum_{\nu=m+1}^{M} \Delta f_\nu = \frac{f_M - f_m}{\lambda_{M-1}} + \sum_{\nu=m+1}^{M-1} (f_\nu - f_m)\left[\frac{1}{\lambda_{\nu-1}} - \frac{1}{\lambda_\nu}\right].$$

Now, in view of IV 2.2 a) & b) we get

IV 2.4 $$\sum_{\nu=m+1}^{M-1} |f_\nu - f_m|\left[\frac{1}{\lambda_{\nu-1}} - \frac{1}{\lambda_\nu}\right] \le 2 \sum_{\nu=m+1}^{M-1} \lambda_{\nu-1}\left[\frac{1}{\lambda_{\nu-1}} - \frac{1}{\lambda_\nu}\right] \le \log \frac{\lambda_{M-1}}{\lambda_m} \quad (*)$$

Thus $\{h_n\}$ converges as $n \to \infty$ whenever $\lambda_\infty < \infty$ and the limit h is clearly integrable.

This given, we set for each $\alpha > 0$

$$h^\alpha = \sum_{\nu=1}^{\infty} \frac{\Delta f_\nu}{\lambda_{\nu-1}} \chi(\lambda_{\nu-1} \le \alpha) = \sum_{\nu=1}^{\infty} \Delta h_\nu \chi(\lambda_{\nu-1} \le \alpha).$$

Now, the following identity holds

$$h^\alpha = h \chi(\lambda_\infty \le \alpha) + \sum_{m=0}^{\infty} \chi(\lambda_{m-1} \le \alpha < \lambda_m) h_m$$

$(*)$ We have used $1 - x \le \log \frac{1}{x}$ for $0 < x < 1$.

where, for convenience we set $\lambda_{-1} = 0$.

Thus for each $\alpha > 0$ we can write

IV 2.5
$$h^{2\alpha} - h^{\alpha} = \sum_{m=0}^{\infty} \chi(\lambda_{m-1} \leq \alpha < \lambda_m) \left(h^{2\alpha} - h_m\right) =$$
$$= \sum_{m=0}^{\infty} \chi(\lambda_{m-1} \leq \alpha < \lambda_m) \left(h^{2\alpha} - h_m^{2\alpha}\right)$$

In the set where $\lambda_{m-1} \leq \alpha < \lambda_m$ either $\lambda_M \leq 2\alpha$ for all M or there is an M such that $\lambda_{M-1} \leq 2\alpha < \lambda_M$ in either case we can easily see that IV 2.4 gives

$$\left|h^{2\alpha} - h^{\alpha}\right| \leq 2 + 2 \log 2 .$$

Thus when $E(|\varphi - \varphi_m| | \mathcal{F}_m) \leq \gamma_m = E(\gamma | \mathcal{F}_m)$, the identity

$$E\left((h^{2\alpha} - h^{\alpha})\varphi\right) = \sum_{m=0}^{\infty} E\left(\chi(\lambda_{m-1} \leq \alpha < \lambda_m)(h^{2\alpha} - h_m^{2\alpha})(\varphi - \varphi_m)\right)$$

yields

$$\left|E\left((h^{2\alpha} - h^{\alpha})\varphi\right)\right| \leq (2 + 2 \log 2) \sum_{m=0}^{\infty} E\left(\chi(\lambda_{m-1} \leq \alpha < \lambda_m) \gamma_m\right) \leq$$
$$\leq (2 + 2 \log 2) E\left(\chi(\alpha < \lambda_{\infty}) \gamma\right) .$$

Note now that for any $B < \infty$

$$\int_0^B E\left(|h^{2\alpha} - h^\alpha||\varphi|\right) d\alpha \leq (2 + 2 \log 2) \, B \, E(|\varphi|).$$

So by Fubini's theorem we can conclude that

IV 2.6 $\quad \left| E\left(\left[\int_0^B (h^{2\alpha} - h^\alpha) \, d\alpha \right] \varphi \right) \right| \leq (2 + 2 \log 2) \, E(\lambda_\infty \gamma).$

Now, if f is bounded and $B \geq \|f\|_\infty$, we can assume (upon replacing λ_{n-1} by $\lambda_{n-1} \wedge B$ is necessary) that $\lambda_\infty \leq B$ as well. But then we have

IV 2.7 $\quad \displaystyle\int_0^B (h^{2\alpha} - h^\alpha) \, d\alpha = \sum_{\nu=1}^\infty \int_0^B \chi\!\left(\frac{\lambda_{\nu-1}}{2} \leq \alpha < \lambda_{\nu-1}\right) d\alpha \, \frac{\Delta f_\nu}{\lambda_{\nu-1}} =$

$$= \frac{1}{2} \sum_{\nu=1}^\infty \Delta f_\nu = \frac{1}{2} f.$$

Here the interchange of summation and integration is justified whenever $\lambda_\infty < \infty$ since the series

$$\sum_{\nu=1}^{\infty} \frac{\Delta f_\nu}{\lambda_{\nu-1}}$$

is boundedly convergent there.

Combining IV 2.7 and IV 2.6, our inequality IV 2.3 immediately follows.

Remark IV 2.1   The inequality IV 2.3 clearly yields

IV 2.8 $$\left|E(f\varphi)\right| \leq (4 + 4 \log 2) \, \|f\|_{\rho_p} \, \|\varphi\|_{\mathcal{K}_q^+} \, ,$$

thus as soon as we show that the bounded functions are dense in ρ_p, IV 2.4 itself can be used to extend $E(f\varphi)$ to all $f \in \rho_p$ and all $\varphi \in \mathcal{K}_q^+$.

But the remarkable fact is that

IV 2.9 $$\lim_{B \to \infty} \left\| f - 2 \int_0^B (h^{2\alpha} - h^\alpha) \, d\alpha \right\|_{\rho_p} = 0 \, .$$

Thus we can simply define

$$E(f\varphi) = \lim_{B \to \infty} 2 \int_0^B E\!\left((h^{2\alpha} - h^\alpha)\varphi\right) d\alpha$$

To show IV 2.8, note that

$$f - 2\int_0^B (h^{2\alpha} - h^\alpha)\, d\alpha = \sum_{\nu=1}^\infty \Delta f_\nu\, \theta_B(\lambda_{\nu-1}) = f^B$$

where

$$\theta_B(\lambda) = 1 - \frac{2}{\lambda}\int_0^B \chi\left(\frac{\lambda}{2} \le \alpha < \lambda\right) d\alpha = 1 - \left(\frac{2B}{\lambda} - 1\right)^+ \wedge 1\ .$$

But the identity

$$f_n^B = f_n\, \theta_B(\lambda_{n-1}) + \sum_{\nu=1}^{n-1} f_\nu \left[\theta_B(\lambda_\nu) - \theta_B(\lambda_{\nu-1})\right],$$

together with the fact that $\theta_B(\lambda)$ is an increasing function of λ, shows that

$$|f_n^B| \le 2\, \lambda_{n-1}\, \theta_B(\lambda_{n-1})$$

thus

$$\|f^B\|_{\mathcal{P}_p} \le 2 \left[E\left(\lambda_\infty^p\, \theta_B^p(\lambda_\infty)\right)\right]^{1/p}\ .$$

Since $\theta_B(\lambda) \leq 1$ and $\theta_B(\lambda) \to 0$ as $B \to \infty$, the dominated convergence theorem finally yields IV 2.8 as asserted.

Remark IV 2.2   If we go back for a moment to Section III 3 and look over the proof of the Burgess Davis decomposition (Theorem III 3.5) we can deduce that every function $f \in \mathcal{H}_p$ ($p \geq 1$) can be written in the form

$$f = h + g$$

where

IV 2.10 $\qquad |\Delta h_\nu| \leq 4\,(f^*_\nu - f^*_{\nu-1}) + 4\,E(f^*_\nu - f^*_{\nu-1} | \mathcal{F}_{\nu-1})$

and

IV 2.11 $\qquad g^*_n \leq 13\,f^*_{n-1} + 4 \sum_{\nu=1}^{n-1} E(f^*_\nu - f^*_{\nu-1} | \mathcal{F}_{\nu-1})\;.$

Now, if a certain $\varphi \in L_1$ satisfies

$$E(|\varphi - \varphi_{n-1}| | \mathcal{F}_n) \leq E(\gamma | \mathcal{F}_n) \qquad \forall\; n \geq 1$$

for a $\gamma \in L_q$, then clearly $\varphi \in \mathcal{K}^-_q$ and in addition we have also

IV 2.12 $$|\Delta\varphi_n| \le E(\gamma|\mathfrak{F}_n) = \gamma_n$$

Because of IV 2.11 g belongs to $\mathcal{P}_p$. Indeed, if we set

$$A_{n-1}(f^*) = \sum_{\nu=1}^{n-1} E(f^*_\nu - f^*_{\nu-1}|\mathfrak{F}_{\nu-1})$$

one of the inequalities proved in Section III 4 yields

IV 2.13 $$E([A_\infty(f^*)]^p) \le p^p E(f^{*p}) .$$

Thus, using IV 2.3 and IV 2.10 we get

$$|E(g\varphi)| \le (4 + 4 \log 2) E([13 f^* + 4 A_\infty(f^*)] \gamma) .$$

On the other hand, IV 2.10 and IV 2.12 give

$$|E(h\varphi)| \le \sum_{\nu=1}^{\infty} E\left(|\Delta h_\nu||\Delta\varphi_\nu|\right) \le$$
$$\le 4 \sum_{\nu=1}^{\infty} E([f^*_\nu - f^*_{\nu-1}] \gamma_\nu) + 4 \sum_{\nu=1}^{\infty} E\left(\gamma_\nu E(f^*_\nu - f^*_{\nu-1}|\mathfrak{F}_{\nu-1})\right) =$$
$$= 4 E(\gamma f^*) + 4 E\left(\gamma A_\infty(f^*)\right) .$$

Putting all this together we finally obtain

$$|E(f\varphi)| \leq (56 + 52 \log 2)\, E\left(\gamma f^*\right) + (20 + 16 \log 2)\, E\left(\gamma A_\infty(f^*)\right).$$

It is clear, in view of IV 2.13 that this inequality would lead to a new proof of Theorem III 5.2 but the constants involved would be considerably larger. At any rate, we can still state the following extension of Theorem III 3.5.

<u>Theorem IV 2.2</u> <u>Every function</u> $f \in \mathcal{H}_p$ $(p \geq 1)$ <u>can be written</u> <u>in the form</u>

$$f = h + g$$

where

$$h \in G_p = \{f: \sum_{\nu=1}^{\infty} |\Delta f_\nu| \in L_p\},$$

$$g \in \mathcal{P}_p$$

and

$$\left\| \sum_{\nu=1}^{\infty} |\Delta h_\nu| \right\|_p \leq (4 + 4p)\, \|f^*\|_p,$$

$$\|g\|_{\mathcal{P}_p} \leq (13 + 4p)\, \|f^*\|_p.$$

IV 3 L_p estimates between f^* and the conditional square function.

Gundy and Burkholder established in [8] also the inequalities

IV 3.1 $$\|f^*\|_p \leq c_p^{(1)} \|\sigma(f)\|_p \quad 0 < p \leq 2,$$

IV 3.2 $$\|\sigma(f)\|_p \leq c_p^{(2)} \|f^*\|_p \quad 2 \leq p \leq \infty. \quad (*)$$

Of course, from the material already presented we can at least derive IV 3.1 for $1 \leq p \leq 2$ and IV 3.2 $\forall\, p \geq 2$. Indeed, if we combine IV 1.4 and II 2.8 we get

$$\|f^*\|_p \leq 2\sqrt{5} \, \|\sigma(f)\|_p \quad 1 \leq p \leq 2,$$

while IV 1.5 and II 1.3 give

$$\|\sigma(f)\|_p \leq p \, \|f^*\|_p \quad 2 \leq p < \infty.$$

But we can do much better here by working directly upon $\sigma(f)$.

(*) See [8] Lemma 2.2 for IV 3.1 and Theorem 5.3 for IV 3.2.

In fact, using our martingale transform techniques we can show that if for some $p > 0$

IV 3.3 $\qquad E([\sigma(f)]^p) = E\left(\left[\sum_{\nu=1}^{\infty} E\left([\Delta f_\nu]^2 \mid \mathfrak{F}_{\nu-1}\right)\right]^p\right) < \infty$

then f can be obtained as a martingale transform of a g in L_2. As a result we are able to show that both IV 3.1 and IV 3.2 can be regarded as simple consequences of the identity

IV 3.4 $\qquad E\left(\sigma^2(g)\right) = \sum_{\nu=1}^{\infty} E\left([\Delta g_\nu]^2\right) = E(g^2).$ (*)

To see this, suppose IV 3.3 holds, and set

IV 3.5 $\qquad g_n = \sum_{\nu=1}^{n} \Delta f_\nu [\sigma_\nu(f)]^{(p-2)/2}$

We then have

(*) Remember, all our functions here are normalized to have conditional expectation with respect to $\mathfrak{F}_0$ equal to zero!

$$\sigma^2(g_n) = \sum_{\nu=1}^{n} E\Big([\Delta f_\nu]^2 \big| \mathfrak{F}_{\nu-1}\Big) [\sigma_\nu(f)]^{p-2} =$$

$$= \sum_{\nu=1}^{n} [\sigma_\nu^2(f) - \sigma_{\nu-1}^2(f)] [\sigma_\nu(f)]^{p-2}$$

Thus, using the elementary inequalities

$$\frac{2}{p}\left(B^p - A^p\right) \geq \left(B^2 - A^2\right) B^{p-2} \quad \text{for} \quad 0 < p \leq 2 ,$$

$$\frac{2}{p}\left(B^p - A^p\right) \leq \left(B^2 - A^2\right) B^{p-2} \quad \text{for} \quad 2 \leq p ,$$

(valid whenever $B \geq A$), we immediately obtain

$$\sigma_n^p(f) \leq \sigma^2(g_n) \leq \frac{2}{p} \sigma_n^p(f) \quad \text{for} \quad 0 < p \leq 2 ,$$

and

$$\frac{2}{p} \sigma_n^p(f) \leq \sigma^2(g_n) \leq \sigma_n^p(f) \quad \text{for} \quad 2 \leq p .$$

We see then that in any case the martingale g_n converges to a function g in L_2, and in view of IV 3.4 we must have

IV 3.6 $\qquad E(\sigma^p(f)) \leq E(g^2) \leq \dfrac{2}{p} E(\sigma^p(f)) \quad 0 < p \leq 2$,

IV 3.7 $\qquad \dfrac{2}{p} E\left(\sigma^p(f)\right) \leq E(g^2) \leq E\left(\sigma^p(f)\right) \quad 2 \leq p$.

Now, to get IV 3.1 we simply observe that

$$f_n = \sum_{\nu=1}^{n} \Delta g_\nu \, [\sigma_\nu(f)]^{1-p/2}$$

$$= g_n \, [\sigma_n(f)]^{1-p/2} - \sum_{\nu=1}^{n-1} g_\nu \left\{ [\sigma_\nu(f)]^{1-p/2} - [\sigma_{\nu-1}(f)]^{1-p/2} \right\} ,$$

thus

$$f_n^* \leq 2 \, g_n^* \, [\sigma_n(f)]^{1-p/2} .$$

Hölder's inequality and Lemma I 4.2 then give

$$E(f_n^{*p}) \leq 2^p \left[E(g_n^{*2}) \right]^{p/2} \left[E(\sigma_n^p(f)) \right]^{1-p/2}$$

$$\leq 2^p \left[4 \, E(g_n^2) \right]^{p/2} \left[E(\sigma_n^p(f)) \right]^{1-p/2}$$

hence, using IV 3.6 we finally obtain

IV 3.8
$$E(f^{*p}) \le 2^{2p} \left(\frac{2}{p}\right)^{p/2} E\left(\sigma^p(f)\right) .$$

The same basic idea yields IV 3.2, although as we shall see the methods of Section III 4 produce a better inequality in this case.

We start by writing IV 3.5 in the form

$$g_n = f_n[\sigma_n(f)]^{(p-2)/2} - \sum_{\nu=1}^{n-1} f_\nu \left\{[\sigma_\nu(f)]^{(p-2)/2} - [\sigma_{\nu-1}(f)]^{(p-2)/2}\right\},$$

thus

$$|g_n| \le 2 f_n^* [\sigma_n(f)]^{(p-2)/2}$$

Hölder's inequality then gives

$$E(g_n^2) \le 4 \left[E(f_n^{*p})\right]^{2/p} [E(\sigma_n^p(f))]^{1-2/p}$$

combining with IV 3.7

$$E(\sigma_n^p(f)) \le \frac{p}{2} 4 \left[E(f_n^{*p})\right]^{2/p} [E(\sigma_n^p(f))]^{1-2/p}$$

It is clear that the integrability of $\sigma_n^p(f)$ is not in question when $f^* \in L_p$ $(p \ge 2)$, so we finally obtain

IV 3.8
$$\left[E(\sigma^p(f))\right]^{1/p} \leq \sqrt{2p} \left[E(f^{*p})\right]^{1/p} .$$

Note, however, that if we start with the identity

$$E\left(\sigma^2(f) - \sigma_n^2(f) \mid \mathcal{F}_n\right) = E\left(f^2 - f_n^2 \mid \mathcal{F}_n\right) \leq E\left(f^2 \mid \mathcal{F}_n\right)$$

and use Theorem III 4.2 with $m(t) = t^{(p-2)/2}$ we get

$$\frac{2}{p} E\left(\sigma^p(f)\right) \leq E\left(f^2 [\sigma(f)]^{(p-2)/2}\right)$$

and Hölder's inequality gives

$$E\left(\sigma^p(f)\right) \leq \frac{p}{2} \left[E(f^p)\right]^{2/p} \left[E\left(\sigma^p(f)\right)\right]^{1-2/p} .$$

This leads to a much better inequality than IV 3.8 .

So we can finally state:

<u>Theorem IV 3.1</u>   <u>The following inequalities hold</u>

IV 3.9
$$E(f^{*p}) \leq 2^{2p} \left(\frac{2}{p}\right)^{p/2} E\left(\sigma^p(f)\right) \quad 0 < p \leq 2 ,$$

IV 3.10
$$E\left(\sigma^p(f)\right) \leq \left(\frac{p}{2}\right)^{p/2} E(f^p) \quad 2 \leq p .$$

IV 4. Relations amongst $\mathcal{P}_p$ spaces and further remarks.

These martingale transform techniques can be used to show that given any p_1 and p_2 the function of $\mathcal{P}_{p_1}$ are martingale transforms of those of $\mathcal{P}_{p_2}$ and viceversa. The same thing can be done regarding the various spaces H_p^- and BMO^+.

We shall indicate here a few of these results and some open problems. We start by working with the spaces $\mathcal{P}_p$.

First of all recall that if $f \in \mathcal{P}_p$ then by definition

IV 4.1
$$\|f\|_{\mathcal{P}_p} = \inf_{\{\lambda_n\}} \left[E(\lambda_\infty^p) \right]^{\frac{1}{p}}$$

where the "inf" is taken over all sequences $\{\lambda_n\}$ satisfying IV 2.2.

Now, it is clear that the "inf" is achieved. This is because if each $\{\lambda_n^{(k)}\}$ satisfies IV 2.2 and $\|\lambda_\infty^{(k)}\|_p \to \|f\|_{\mathcal{P}_p}$ then if we set

$$\lambda_n = \inf_k \lambda_n^{(k)}$$

$\{\lambda_n\}$ will not only satisfy the conditions in IV 2.2 but also

IV 4.2
$$\|f\|_{\mathcal{P}_p} = \left[E(\lambda_\infty^p) \right]^{1/p} .$$

For convenience of exposition let us then agree that when we write $\{\lambda_n(f)\}$ we implicitly assume that $\{\lambda_n(f)\}$ is one such optimal sequence, that is one which obeys IV 2.2 and IV 4.2 .

Our first result can be stated as follows

<u>Theorem IV 4.1</u>   <u>Let</u> $p_1 < p_2$ <u>and</u> $f \in \mathcal{P}_{p_1}$ <u>then the function</u>

IV 4.3 $$g = \sum_{\nu=1}^{\infty} \frac{\Delta f_\nu}{\left[\lambda_{\nu-1}(f)\right]^{1-p_1/p_2}}$$

<u>is in</u> $\mathcal{P}_{p_2}$ <u>and</u>

IV 4.4 $$1/2 \, \|f\|_{\mathcal{P}_{p_1}}^{p_1/p_2} \leq \|g\|_{\mathcal{P}_{p_2}} \leq p_2/p_1 \, \|f\|_{\mathcal{P}_{p_1}}^{p_1/p_2}$$

Proof. For convenience let $\alpha = 1 - p_1/p_2$, we have

$$g_n = \frac{f_n}{[\lambda_{n-1}(f)]^\alpha} + \sum_{\nu=1}^{n-1} f_\nu \left[\frac{1}{[\lambda_{\nu-1}(f)]^\alpha} - \frac{1}{[\lambda_\nu(f)]^\alpha} \right] .$$

Thus

$$|g_n| \leq \left[\lambda_{n-1}(f)\right]^{1-\alpha} + \sum_{\nu=1}^{n-1} \lambda_{\nu-1}(f) \left[\frac{1}{[\lambda_{\nu-1}(f)]^\alpha} - \frac{1}{[\lambda_\nu(f)]^\alpha}\right]$$

Now, if we agree to let $\lambda_{-1}(f) = 0$, this inequality can be manipulated into

$$|g_n| \leq \sum_{\nu=1}^n \frac{[\lambda_{\nu-1}(f) - \lambda_{\nu-2}(f)]}{[\lambda_{\nu-1}(f)]^\alpha} \leq \int_0^{\lambda_{n-1}(f)} \frac{d\lambda}{\lambda^\alpha} = \frac{p_2}{p_1} \left[\lambda_{n-1}(f)\right]^{p_1/p_2},$$

from which the right half of IV 4.4 immediately follows.

To prove the left half, note that

$$f_n = \sum_{\nu=1}^n \Delta g_\nu \, \lambda_{\nu-1}^\alpha(f) = g_n[\lambda_{n-1}(f)]^\alpha - \sum_{\nu=1}^{n-1} g_\nu \left([\lambda_\nu(f)]^\alpha - [\lambda_{\nu-1}(f)]^\alpha\right),$$

so

$$|f_n| \leq 2 \, \lambda_{n-1}(g) \, [\lambda_{n-1}(f)]^\alpha .$$

Thus

$$\|f\|_{p_1}^{p_1} \leq 2^{p_1} E\left(\lambda_\infty^{p_1}(g) \, [\lambda_\infty(f)]^{\alpha p_1}\right)$$

and using the definition of $\lambda_{n-1}(f)$ plus Hölder's inequality

$$E\left([\lambda_\infty(f)]^{p_1}\right) \leq 2^{p_1} \left[E\left([\lambda_\infty(q)]^{p_2}\right)\right]^{p_1/p_2} \left[E\left([\lambda_\infty(f)]^{p_1}\right)\right]^{1-p_1/p_2}.$$

making the appropriate cancellation we get

$$\|f\|_{\mathcal{P}_{p_1}}^{p_1/p_2} \leq 2 \, \|g\|_{\mathcal{P}_{p_2}}$$

as asserted.

Theorem IV 4.1 has the following corollary:

<u>Theorem IV 4.2.</u> <u>If</u> $f \in \mathcal{P}_{p_1}$ <u>and</u> $p_2 > p_1$ <u>then there is a function</u> $g \in \mathcal{P}_{p_2}$ <u>such that</u>

$$f = \sum_{\nu=1}^{\infty} \Delta g_\nu \, T_{\nu-1}$$

<u>where</u> $\{T_\nu\}$ <u>is adapted, non-decreasing and</u>

a) $\|g\|_{\rho_{p_2}} \leq \dfrac{p_2}{p_1} \|f\|_{\rho_{p_1}}^{p_1/p_2}$,

b) $E(T_\infty^\alpha) \leq \|f\|_{\rho_{p_1}}^{p_1}$, $\alpha = \dfrac{p_2 \, p_1}{p_2 - p_1}$.

In a completely analogous manner we can prove the following converse result:

<u>Theorem IV 4.3.</u> <u>If</u> $g \in \rho_{p_2}$ <u>and</u> $p_1 < p_2$ <u>then there is a function</u> $f \in \rho_{p_1}$ <u>such that</u>

$$g = \sum_{\nu=1}^{\infty} \dfrac{\Delta f_\nu}{T_{\nu-1}}$$

<u>where</u> $\{T_\nu\}$ <u>is adapted, non-decreasing and</u>

a) $\|f\|_{\rho_{p_1}} \leq 2 \, \|g\|_{\rho_{p_2}}^{p_2/p_1}$,

b) $E(T_\infty^\alpha) \leq \|g\|_{\rho_{p_2}}$, $\alpha = \dfrac{p_1 \, p_2}{p_2 - p_1}$

For the proof set

$$f = \sum_{\nu=1}^{\infty} \Delta g_\nu \, T_{\nu-1}$$

where $T_{\nu-1} = [\lambda_{\nu-1}(g)]^{-1+p_2/p_1}$.

Remark IV 4.1 Some interesting questions arise if we go to the extreme case $p_2 = \infty$. For simplicity let us work with $p_1 = 2$. One might be tempted to conjecture in this case that every $f \in \mathcal{P}_2$ can be written as a martingale transform of a <u>bounded</u> function.

Indeed, if $\|h\|_\infty < \infty$ and $\{T_\nu\}$ is an adapted, non-decreasing sequence, the function

IV 4.5
$$f = \sum_{\nu=1}^{\infty} \Delta h_\nu \, T_{\nu-1}$$

is in $\mathcal{P}_2$ as soon as $E(T_\infty^2) < \infty$.

This is clear since the identity

$$f_n = h_n \, T_{n-1} - \sum_{\nu=1}^{n-1} h_\nu [T_\nu - T_{\nu-1}]$$

gives again

$$|f_n| \leq 2 \, \|h\|_\infty \, T_{n-1} \; .$$

But our methods here fail to give the converse result, namely that every $f \in \mathcal{P}_2$ can be written in this form. It is interesting to note however that we can always obtain IV 4.5 with $h \in$ BMO !

To show this simply set

$$h = \sum_{\nu=1}^{\infty} \frac{\Delta f_\nu}{T_{\nu-1}}$$

where

IV 4.6
$$T_\nu = \max_{0 \leq \mu \leq \nu} E \, \lambda_\infty(f) \big| \mathcal{F}_\mu \; .$$

We then have

$$h_n - h_{\mu-1} = \frac{f_n - f_{\mu-1}}{T_{n-1}} + \sum_{\nu=\mu}^{n-1} \left(f_\nu - f_{\mu-1}\right) \left[\frac{1}{T_{\nu-1}} - \frac{1}{T_\nu}\right].$$

Thus

$$|h_n - h_{\mu-1}| \leq 2 \frac{\lambda_{n-1}(f)}{T_{n-1}} + 2 \sum_{\nu=\mu}^{n-1} \lambda_{\nu-1}(f) \left[\frac{1}{T_{\nu-1}} - \frac{1}{T_\nu}\right] =$$

$$= 2 \frac{\lambda_{\mu-1}(f)}{T_{\mu-1}} + \sum_{\nu=\mu+1}^{n} \frac{\lambda_{\nu-1}(f) - \lambda_{\nu-2}(f)}{T_{\nu-1}}$$

Now, from IV 4.6 we easily get

IV 4.7

a) $\lambda_{\nu-1}(f) \leq T_{\nu-1}$

b) $1 \leq E(\lambda_\infty(f)|\mathcal{F}_{\nu-1}) \, E\left(\frac{1}{\lambda_\infty(f)} \Big| \mathcal{F}_{\nu-1}\right) \leq T_{\nu-1} \, E\left(\frac{1}{\lambda_\infty(f)} \Big| \mathcal{F}_{\nu-1}\right).$

So we finally have

$$E\left(|h_n - h_{\mu-1}| \Big| \mathcal{F}_\mu\right) \leq 2 + 2 \, E\left(\sum_{\nu=\mu+1}^{n} \left(\lambda_{\nu-1}(f) - \lambda_{\nu-2}(f)\right) E\left(\frac{1}{\lambda_\infty(f)} \Big| \mathcal{F}_\mu\right)\right)$$

$$= 2 + 2 \, E\left(\sum_{\nu=\mu+1}^{n} \left(\frac{\lambda_{\nu-1}(f) - \lambda_{\nu-2}(f)}{\lambda_\infty(f)}\right) \Big| \mathcal{F}_\mu\right) \leq 4 \quad ,$$

and we can state

Theorem IV 4.3 *If* $f \in \mathcal{P}_2$ *then*

$$f = \sum_{\nu=1}^{\infty} \Delta h_\nu \, T_{\nu-1}$$

where $h \in BMO$,

$$\|h\|_{BMO_1} \leq 4 \quad,$$

__and__ $\{T_\nu\}$ is an adapted, non-decreasing sequence such that

$$E(T_\infty^2) \leq 4 \; E\left(\lambda_\infty^2(f)\right) = 4 \; \|f\|_{\mathcal{P}_2}^2 \;.$$

__Remark IV 4.2__ If we work in a similar manner with

$$\mathcal{H}_2 = \{f: \; E\left(\sigma^2(f)\right) < \infty\} = L_2$$

and

$$BMO_2^+ = \{h: \; \sup_n \; \|E(|h - h_n|^2 |\mathcal{F}_n)\|_\infty < \infty\}$$

we get a much more satisfactory result, namely:

__Theorem IV 4.4__ A function f __is in__ L_2 __if and only if it can be written in the form__

IV 4.8
$$f = \sum_{\nu=1}^{\infty} \Delta h_\nu \, T_{\nu-1}$$

where $\{T_\nu\}$ is an adapted non-decreasing process such that $E(T_\infty^2) < \infty$ and $h \in BMO_2^+$.

Proof. Suppose IV 4.8 holds with

$$E(|h - h_n|^2 | \mathcal{F}_n) \leq B^2$$

then

$$\sigma^2(f) = \sum_{\nu=1}^{\infty} E(|\Delta h_\nu|^2 | \mathcal{F}_{\nu-1}) \, T_{\nu-1}^2 \, .$$

Thus, if we set for convenience $T_{\nu-1}^2 = \sum_{\mu=0}^{\nu-1} p_\mu$ we get

$$E(\sigma^2(f)) = \sum_{\nu=1}^{\infty} \sum_{\mu=0}^{\nu-1} E(p_\mu [\Delta h_\nu]^2) = \sum_{\mu=0}^{\infty} \sum_{\nu=\mu+1}^{\infty} E(p_\mu [\Delta h_\nu]^2) =$$

$$= \sum_{\mu=0}^{\infty} E\left(p_\mu E\left(|h - h_\mu|^2 | \mathcal{F}_\mu\right)\right) \leq B^2 \, E(T_\infty^2) \, .$$

That shows $f \in L_2$.

To show the converse set

$$h = \sum_{\nu=1}^{\infty} \frac{\Delta f_\nu}{T_{\nu-1}}$$

where

$$T_\nu = \max_{0 \leq \mu \leq \nu} E\left(\sigma(f) \mid \mathcal{F}_\mu\right).$$

We then have

$$E\left(|h_n - h_\mu|^2 \mid \mathcal{F}_\mu\right) = \sum_{\nu=\mu+1}^{n} E\left(\frac{[\Delta f_\nu]^2}{T_{\nu-1}^2} \mid \mathcal{F}_\mu\right) = \sum_{\nu=\mu+1}^{n} E\left(\frac{\sigma_\nu^2(f) - \sigma_{\nu-1}^2(f)}{T_{\nu-1}^2} \mid \mathcal{F}_\mu\right).$$

Now,

$$1 \leq E\left(\sigma(f) \mid \mathcal{F}_{\nu-1}\right) E\left(\frac{1}{\sigma(f)} \mid \mathcal{F}_{\nu-1}\right) \leq T_{\nu-1} E\left(\frac{1}{\sigma(f)} \mid \mathcal{F}_{\nu-1}\right).$$

So

$$\frac{1}{T_{\nu-1}^2} \leq E\left(\frac{1}{\sigma^2(f)} \mid \mathcal{F}_{\nu-1}\right)$$

and we derive

$$E\left(|h_n - h_\mu|^2 | \mathcal{F}_\mu\right) \le \sum_{\nu=\mu+1}^{n} E\left(\frac{\sigma_\nu^2(f) - \sigma_{\nu-1}^2(f)}{\sigma^2(f)} | \mathcal{F}_\mu\right)$$

While at the same time we have

$$E(T_\infty^2) \le 4 E\left(\sigma^2(f)\right) = 4 \|f\|_2^2 .$$

<u>Remark IV 4.3</u> If we go through the second half of the proof of Theorem IV 4.4 we see that the hypothesis that $\sigma(f) \in L_2$ does not play any role in the proof that $h \in BMO_2^+$. Indeed, the proof needs only $\sigma(f) \in L_1$. If $\sigma(f) \in L_p$ $(p > 1)$ then we get as well $\|T_\infty\|_p \le q \|\sigma(f)\|_p$. So we can always represent an $f \in \bar{\mathcal{H}}_p$ $(p > 1)$ in the form IV 4.8 with $\|h\|_{BMO_2^+} \le 1$ and $\|T_\infty\|_p < \infty$. However, the proof fails for $p = 1$. It is not clear what remains true in that case. In this connection, it is interesting to observe that if

IV 4.9
$$E\left(\sqrt{\sigma^2(h) - \sigma_\nu^2(h)} | \mathcal{F}_\nu\right) \le B$$

and

$$f = \sum_{\nu=1}^{\infty} \Delta h_\nu \, T_{\nu-1}$$

with $E(T_\infty) < \infty$ then $f \in H_1^-$!

This is rather easy to show. On the other hand since the inequality IV 3.9 is true with <u>conditioning</u> as well we necessarily have

$$E(|h - h_\nu| \,|\, \mathfrak{F}_\nu) \leq 4\sqrt{2}\, E\!\left(\sigma(h - h_\nu) \,|\, \mathfrak{F}_\nu\right)$$
$$= 4\sqrt{2}\, E\!\left(\sqrt{\sigma^2(h) - \sigma_\nu^2(h)} \,\big|\, \mathfrak{F}_\nu\right).$$

Of course Schwarz's inequality gives

$$E\!\left(\sqrt{\sigma^2(h) - \sigma_\nu^2(h)} \,\big|\, \mathfrak{F}_\nu\right) \leq \sqrt{E\!\left(|h - h_\nu|^2 \,\big|\, \mathfrak{F}_\nu\right)}$$

So we see that the condition in IV 4.9 is sort of in between $h \in BMO_1^+$ and $h \in BMO_2^+$. This suggests that perhaps Theorem IV 4.4 does not extend all the way down to H_1^-.

<u>Remark IV 4.4</u>   We shall close by showing that these martingale transform techniques can be used to prove convergence results as well.

Indeed, let $\{f_n\}$ be a $\{\mathfrak{F}_n\}$-martingale and suppose that the martingale transform

$$g_n = \sum_{\nu=1}^{n} \frac{\Delta f_\nu}{\lambda_{\nu-1}}$$

$(\lambda_0 \leq \lambda_1 \leq \ldots \leq \lambda_n \leq \ldots \leq \lambda_\infty$, $\mathfrak{F}(\lambda_\nu) \subset \mathfrak{F}_\nu)$ converges to a function g in L_2 then f_n converges a.s. in the set where $\lambda_\infty < \infty$. This follows immediately from the identity

$$f_n = \lambda_{n-1} g_n - \sum_{\nu=1}^{n-1} g_\nu (\lambda_\nu - \lambda_{\nu-1})$$

But now, assuming that the functions

$$\sigma^2(f_n) = \sum_{\nu=1}^{n} E([\Delta f_\nu]^2 | \mathfrak{F}_{\nu-1})$$

do exist, we can set

$$\lambda_{\nu-1} = \sigma^2(f_\nu) + 1 \quad (\lambda_{-1} = 1)$$

and derive

$$\sigma^2(g_n) = \sum_{\nu=1}^{n} \frac{1}{\lambda_{\nu-1}^2} E([\Delta f_\nu]^2 | \mathfrak{F}_{\nu-1}) = \sum_{\nu=1}^{n} \frac{1}{\lambda_{\nu-1}^2} [\lambda_{\nu-1} - \lambda_{\nu-2}] \leq 1$$

So clearly in this case g_n is an L_2 bounded martingale and we thus derive the result that in the set where $\sup_n \sigma(f_n) < \infty$ the martingale f_n converges almost surely.

NOTATION AND BASIC DEFINITIONS

We work with a fixed probability space $(\Omega, \mathcal{F}, P)$ and an increasing sequence of σ-fields $\{\mathcal{F}_n\}$ such that $\bigvee_{n=1}^{\infty} \mathcal{F}_n = \mathcal{F}$. In these notes $\{\lambda_n\}$ " adapted " means $\mathcal{F}(\lambda_n) \subset \mathcal{F}_n \quad \forall n$.

If f is integrable, we set $f_n = E(f|\mathcal{F}_n)$ and usually assume $f_0 = 0$. We also set $\Delta f_n = f_n - f_{n-1}$.

$$S_n(f) = \sqrt{\sum_{\nu=1}^{n} [\Delta f_\nu]^2},$$

$$\sigma_n(f) = \sqrt{\sum_{\nu=1}^{n} E([\Delta f_\nu]^2 | \mathcal{F}_{\nu-1})},$$

$$f_n^* = \max_{0 \leq \nu \leq n} |f_\nu|.$$

The basic functions we shall work with are:

<u>the square function</u> $\qquad S(f) = \lim_{n \to \infty} S_n(f),$

<u>the conditioned square function</u> $\qquad \sigma(f) = \lim_{n \to \infty} \sigma_n(f),$

<u>the maximal function</u> $\qquad f^* = \lim_{n \to \infty} f_n^*.$

A <u>potential</u> is a non-negative supermartingale Q_n which can be written in the form $Q_n = E(A_\infty - A_{n-1}|\mathcal{F}_n)$ where $A_n \uparrow A_\infty$. If $\{A_n\}$ is adapted and $A_0 = 0$ then $\{A_n\}$ is called the "canonical" process associated with Q_n.

If $\{R_n\}$ is a given process, an integrable function $\gamma \geq 0$ is a "majorant" of R_n if and only if

$$|R_n| \leq E(\gamma|\mathcal{F}_n) \quad \forall n .$$

We shall deal with the following spaces:

$$\mathcal{H}_p = \{f: E(S^p(f)) < \infty\} ,$$

$$\mathcal{H}_p^- = \{f: E(\sigma^p(f)) < \infty\} ,$$

$$G_p = \{f: \sum_{\nu=1}^{\infty} |\Delta f_\nu| \in L_p\} ,$$

$$BMO = \{f: \sup_n \|E(|f - f_{n-1}|^2|\mathcal{F}_n)\|_\infty < \infty\} ,$$

$$BMO_p^+ = \{f: \sup_n \|E(|f - f_n|^p|\mathcal{F}_n)\|_\infty < \infty\} ,$$

and the following three spaces $\mathcal{K}_p$, $\mathcal{K}_p^+$ and $\mathcal{P}_p$ whose definition is a bit more elaborate:

NOTATION AND BASIC DEFINITIONS

For $\mathcal{K}_p$ we let

$$\Gamma_f = \left\{\gamma: E(|f - f_{n-1}|^2|\mathcal{F}_n) \leq E(\gamma^2|\mathcal{F}_n)\right\}$$

and set

$$\mathcal{K}_p = \left\{f: \exists\, \gamma \in \Gamma_f \ni \|\gamma\|_p < \infty\right\} \quad (p \geq 2)$$

For $\mathcal{K}_p^+$ we let

$$\Gamma_f^+ = \left\{\gamma: E(|f - f_n||\mathcal{F}_n) \leq E(\gamma|\mathcal{F}_n)\right\}$$

and set

$$\mathcal{K}_p^+ = \{f: \exists\, \gamma \in \Gamma_f^+ \ni \|\gamma\|_p < \infty\} \quad (p \geq 1).$$

For $\mathcal{P}_p$ (L_p-predictable functions) let us say that "$\{\lambda_n\}$ is N.D.A" $\Leftrightarrow$ it is non-decreasing and adapted. This given we set

$$\mathcal{P}_p = \{f: |f_n| \leq \lambda_{n-1}, \text{ where } \{\lambda_n\} \text{ is N.D.A. and } \lambda_\infty \in L_p\}.$$

~~~~~~~~~~

Corresponding to these spaces we have the following norms:

$$\|f\|_{H_p} = \|S(f)\|_p$$

$$\|f\|_{H_p^-} = \|\sigma(f)\|_p$$

$$\|f\|_{\widetilde{G}_p} = \left\|\sum_{\nu=1}^{\infty} |\Delta f_\nu|\right\|_p$$

$$\|f\|_{BMO_1} = \sup_n \|E(|f - f_{n-1}| \mid \mathcal{F}_n)\|_\infty ,$$

$$\|f\|_{BMO_2} = \sup_n \left\|\sqrt{E(|f - f_{n-1}|^2 \mid \mathcal{F}_n)}\right\|_\infty ,$$

$$\|f\|_{BMO_p^+} = \sup_n \left\|[E(|f - f_n|^p \mid \mathcal{F}_n)]^{1/p}\right\|_\infty$$

$$\|f\|_{\mathcal{K}_p} = \inf_{\gamma \in \Gamma_f} \|\gamma\|_p$$

$$\|f\|_{\mathcal{K}_p^+} = \inf_{\gamma \in \Gamma_f^+} \|\gamma\|_p$$

$$\|f\|_{\mathcal{P}_p} = \inf \|\lambda_\infty\|_p$$

where the last "inf" is taken over all $\{\lambda_n\}$ N.D.A. such that $|f_n| \leq \lambda_{n-1}$ $\forall n$.

~~~~~~~~~~

INDEX TO THE INEQUALITIES

Fefferman's inequality

$$|E(f\varphi)| \leq \sqrt{2} \; \|f\|_{H_1} \; \|\varphi\|_{BMO_2} \qquad \text{p. 8}$$

~~~~~~~~~~

Extention to $1 \leq p \leq 2$ of the above

$$|E(f\varphi)| \leq \sqrt{2/p} \; \|f\|_{H_p} \; \|f\|_{H_q} \qquad \text{p. 8}$$

where $q = \dfrac{p}{p-1}$, $1 \leq p \leq 2$.

~~~~~~~~~~

Herz's generalization of Fefferman's inequality

$$|E(f\varphi)| \leq 12 \; \|f\|_{P_1} \; \|\varphi\|_{BMO_1^+} \qquad \text{p. 131}$$

~~~~~~~~~~

Extension to $p \geq 1$ of the above

$$|E(f\varphi)| \leq (4 + 4 \log 2) \|f\|_{P_p} \|\varphi\|_{\mathcal{K}_q^+} , \qquad \text{p. 137}$$

$q = \dfrac{p}{p-1}$.

~~~~~~~~~~

Inequalities between $S(f)$ and f^* (Burkholder-Gundy for $p \geq 1$, B. Davis for $p = 1$)

$$\|f^*\|_p \leq \sqrt{10\,p} \; \|S(f)\|_p \qquad 1 \leq p \leq 2 \qquad \text{p. 37}$$

$$\|f^*\|_p \leq 5\,pq \; \|S(f)\|_p \qquad 2 \leq p \qquad \text{p. 41}$$

$$\|S(f)\|_p \leq 5 \; \|f^*\|_p \qquad 1 \leq p \leq 2 \qquad \text{p. 40}$$

$$\|S(f)\|_p \leq \sqrt{2p} \; \|f^*\|_p \qquad 2 \leq p \qquad \text{p. 28}$$

~~~~~~~~~~

Between $S(f)$ and $\sigma(f)$

p. 126

$$\|S(f)\|_p \leq \sqrt{2/p} \; \|\sigma(f)\|_p \qquad 0 < p \leq 2$$

$$\|\sigma(f)\|_p \leq \sqrt{p/2} \; \|S(f)\|_p \qquad 2 \leq p$$

~~~~~~~~~~

INDEX TO THE INEQUALITIES

Between $|f|$, f^* and $\sigma(f)$ p. 147

$$E(|f|^p) \le E(f^{*p}) \le 2^{2p} \left(\frac{2}{p}\right)^{p/2} E(\sigma^p(f)) \qquad 0 < p \le 2$$

$$\|\sigma(f)\|_p \le \sqrt{p/2} \; \|f\|_p \qquad 2 \le p \; .$$

~~~~~~~~~~

Between f^* and a majorant γ of $E(|f - f_{n-1}| \, | \, \mathcal{F}_n)$

$$\|f^*\|_p \le 3 \, pqe \, \|\gamma\|_p \qquad 1 < p < \infty \qquad\qquad \text{p. 116}$$

~~~~~~~~~~

Between various norms

$$\|f\|_{\mathcal{H}_p} \le \|f\|_{\mathcal{H}_p} \le \sqrt{p/2} \; \|f\|_{\mathcal{H}_p} \qquad (p \ge 2) \qquad \text{p. 31}$$

$$\|\varphi\|_{BMO_2} \le 8 \sqrt{2} \; \|\varphi\|_{BMO_1} \qquad\qquad\qquad \text{p. 80}$$

$$\frac{1}{\sqrt{2}} \|f\|_{\mathcal{H}_1} \le \|f\|_{\mathcal{H}_1^-} \le 2 \|f\|_{\mathcal{P}_1} \qquad\qquad\qquad \text{p. 130}$$

~~~~~~~~~~

For regular martingales (i.e. $|\Delta f_n| \leq c\, E(|\Delta f_n| | \mathfrak{F}_{n-1}))$

$$\|f\|_{\mathcal{P}_1} \leq (1 + 3c)\, E(f^*) .$$ p. 96

~~~~~~~~

Burkholder's weak-L_1 inequality for $S(f)$

$$P[S(f) > \lambda] \leq \frac{6}{\lambda} E(|f|)$$ p. 61

~~~~~~~~

Between the square function of a supermartingale Q_n and a majorant γ

$$E\left(S^2(Q_n)\right) \leq 4\, E(\gamma^2)$$ p. 63

~~~~~~~~

For BMO functions

$$E\left(e^{|\varphi - \varphi_{n-1}|} | \mathfrak{F}_n\right) \leq \frac{1}{1 - 8\|\varphi\|_{BMO_1}}$$ p. 65

$$E\left(e^{S^2(f)}\right) \leq \frac{1}{1 - \|\varphi\|_{BMO_2}}$$ p. 69

~~~~~~~~

INDEX TO THE INEQUALITIES

For non-negative martingales

$$E\left(e^{\alpha[S(f)/f^*]^2}\right) \leq \frac{e^{2\alpha}}{1-6\alpha} \qquad \text{p. 71}$$

~~~~~~~~~

Burkholder, Gundy & Davis

For $\Phi(u)$ convex such that

$$p = \sup_u \frac{u\,\Phi'(u)}{\Phi(u)} < \infty$$

$$E\left(\Phi\left(\sum_{\nu=1}^{\infty} E(\epsilon_\nu|\mathcal{F}_\nu)\right)\right) \leq p^{p+1}\, E\left(\Phi\left(\sum_{\nu=1}^{\infty} \epsilon_\nu\right)\right) \qquad \text{p. 107}$$

where $\{\epsilon_\nu\}$ is any sequence of non-negative integrable functions

~~~~~~~~~

For $\Phi(u) = u^p$

$$p^p\, E\left(\left[\sum_{\nu=1}^{\infty} \epsilon_\nu\right]^p\right) \leq E\left(\left[\sum_{\nu=1}^{\infty} E(\epsilon_\nu|\mathcal{F}_\nu)\right]^p\right) \quad 0 < p \leq 1,$$

$$E\left(\left[\sum_{\nu=1}^{\infty} E(\epsilon_\nu|\mathcal{F}_\nu)\right]^p\right) \leq p^p\, E\left(\left[\sum_{\nu=1}^{\infty} \epsilon_\nu\right]^p\right) \quad 1 \leq p,$$

~~~~~~~~~

For adapted sequences $\{A_n\}$ such that $\forall\, n \geq \nu$

$$\|E(|A_n - A_{\nu-1}|\,|\,\mathcal{F}_\nu)\|_\infty \leq B < \infty$$

$$E\left(\exp\left[\max_{\nu \leq m \leq n} |A_m - A_{\nu-1}|\right]\,\Big|\,\mathcal{F}_\nu\right) \leq \frac{1}{1-8B} \qquad \text{p. 79}$$

~~~~~~~~~~

NOTEWORTHY RESULTS AND REMARKS

<u>First representation theorem for</u> BMO   p. 16 & 22

$\varphi \in $ BMO $\Leftrightarrow$ there exists a sequence $\{\sigma_n\}$ with

$$\left\| \sum_{n=1}^{\infty} \sigma_n^2 \right\|_{\infty} < \infty$$

such that

$$\varphi = \sum_{\nu=1}^{\infty} [E(\sigma_\nu | \mathcal{F}_\nu) - E(\sigma_\nu | \mathcal{F}_{\nu-1})]$$

It is also shown that in this case

$$\|\varphi\|_{BMO_2} \leq 2 \left\| \sqrt{\sum_{n=1}^{\infty} \sigma_n^2} \right\|_{\infty}.$$

~~~~~~~~~~

<u>An extension of the above for</u> $\mathcal{H}_p (p \geq 2)$ p. 16 & 22

(See Sections I.4 and I.5, especially Theorem I.5.2)

~~~~~~~~~~

173

Second representation theorem for BMO   p. 49

$\varphi \in BMO \Leftrightarrow$ there exists a sequence $\{\varepsilon_n\}$ with

$$\left\|\sum_{n=1}^{\infty} |\varepsilon_n|\right\|_{\infty} < \infty$$

such that

$$\left\|\varphi - \sum_{\nu=1}^{\infty} E(\varepsilon_\nu|\mathcal{F}_\nu)\right\|_{\infty} < \infty$$

It is also shown that

$$\left\|\sum_{\nu=1}^{\infty} E(\varepsilon_\nu|\mathcal{F}_\nu)\right\|_{BMO_2} \leq \sqrt{5} \left\|\sum_{\nu=1}^{\infty} \varepsilon_\nu\right\|_{\infty} \qquad \text{p. 48}$$

~~~~~~~~~~

A sufficient condition for BMO

If $\{\theta_\nu\}$ is adapted and $\theta^* = \sup_n |\theta_n|$ then

$$\varphi = \sum_{\nu=1}^{\infty} \theta_{\nu-1} \left(E\left(\frac{1}{\theta^*}\big|\mathcal{F}_\nu\right) - E\left(\frac{1}{\theta^*}\big|\mathcal{F}_{\nu-1}\right) \right)$$

is BMO and

NOTEWORTHY RESULTS AND REMARKS

$$\|\varphi\|_{BMO_2} \leq \sqrt{2} \qquad \text{p. 40}$$

This result implies the inequality $E(S(f)) \leq (2 + \sqrt{5}) E(f^*)$, of Burgess Davis.

~~~~~~~~~~

For the relation between the 2nd representation theorem and the B. Davis inequalities. See Remark II 4.1 p. 56.

~~~~~~~~~~

The basic theorem for BMO sequences

If $\{A_n\}$ is adapted and

$$\|E(|A_n - A_{\nu-1}| \,|\, \mathcal{F}_\nu)\|_\infty \leq B \qquad \forall\ n \geq \nu$$

then $A_n^* = \max_{\nu \leq n} |A_\nu|$ satisfies

$$\|E(A_n^* - A_{\nu-1}^* \,|\, \mathcal{F}_\nu)\|_\infty \leq 8B \qquad \forall\ n \geq \nu \qquad \text{p. 15}$$

~~~~~~~~~~

An extension to $p > 1$ of the above is as follows

If $\{A_n\}$ is adapted and

$$E(|A_n - A_{\nu-1}| \,|\, \mathfrak{F}_\nu) \leq E(\gamma | \mathfrak{F}_\nu) \qquad \forall \; n \geq \nu$$

then

$$E\left(A_n^* - A_{\nu-1}^* \,\big|\, \mathfrak{F}_\nu\right) \leq (3\,pqe)^p \, E(\gamma^p | \mathfrak{F}_\nu) \qquad \forall \; n \geq \nu .$$

(see Theorem III 5.1 p. 116)

~~~~~~~~~~

The John-Nirenberg Theorem is obtained in the following form:

$$E\left(e^{\sup_{m \geq \nu} |\varphi_m - \varphi_{\nu-1}|} \,\bigg|\, \mathfrak{F}_\nu \right) \leq \frac{1}{1 - 8\|\varphi\|_{BMO_1}} \qquad \text{p. 79}$$

~~~~~~~~~~

The Burgess Davis decomposition for $\mathcal{H}_1$

Every $f \in \mathcal{H}_1$ can be written in the form

$$f = h + g \quad .$$

NOTEWORTHY RESULTS AND REMARKS

where

$$\|h\|_{G_1} \leq 8 \, E(f^*) \, ,$$

$$\|g\|_{\mathcal{P}_1} \leq 17 \, E(f^*) \qquad \text{p. 91}$$

~~~~~~

<u>An extention to</u> $p > 1$ <u>of the above</u> (see Remark IV 2.2 p. 139).

~~~~~~

<u>For regular sequences</u> $\{\mathfrak{F}_n\}$

$$\mathcal{P}_1 = \bar{\mathcal{H}}_1 = \mathcal{H}_1 \qquad \text{p. 96}$$

~~~~~~

<u>Gundy</u> L_∞-<u>regularity</u> $\Rightarrow$ <u>regularity</u> $(|\Delta f_n| \leq cE(|\Delta f_n| | \mathfrak{F}_{n-1}))$

(See remark III 3.3 p. 88).

~~~~~~

<u>The basic inequality for submartingales</u>

If $\{p_n\}$ is a non-negative submartingale, $p_n^* = \max\limits_{0 \leq \nu \leq n} p_\nu$ and $m(t) \uparrow$ in $[0, \infty)$ then

$$E\left(\int_0^{p_n^*} t\, dm(t)\right) \leq E\left(p_n\, m(p_n^*)\right). \qquad \text{p. 82}$$

~~~~~~~~~~

The basic inequality for the canonical process of a potential

If $E(A_\infty - A_{\nu-1}|\mathcal{F}_\nu) \leq E(\gamma|\mathcal{F}_\nu)$, $\{A_n\}$ adapted $A_n \uparrow A_\infty$ and $A_0 = 0$ then if $m(f) \uparrow$ in $[0, \infty)$

$$E\left(\int_0^{A_n} m(t)\, dt\right) \leq E\left(\gamma\, m(A_n)\right). \qquad \text{p. 102}$$

~~~~~~~~~~

The reverse $L \log L$ result

The following extension of this result of Stein-Burkholder and Gundy is given in p. 88.

If $0 \leq f \in \mathcal{P}_1$ and $\mathcal{F}_0$ is trivial

$$E(f \log^+ f) \leq E(f) \log^+ E(f) + \frac{E(f)}{e} + \|f\|_{\mathcal{P}_1}.$$

~~~~~~~~~~

NOTEWORTHY RESULTS AND REMARKS   179

The $\wp_p$ spaces are all martingale transforms of each other

    See Theorems IV 4.2 and IV 4.3   p.

~~~~~~~~~~

Relation between $\wp_2$ and BMO

    See Remark IV 4.1 Theorem IV 4.4   p.

~~~~~~~~~~

Representation of L_2 functions in terms of BMO functions

    See Remark IV 4.2 Theorem IV 4.5   p.

~~~~~~~~~~

Burkholder's convergence theorem for martingale transforms

    If $\{f_n\}$ is an L_1-bounded martingale and $\{\lambda_n\}$ is adapted then the martingale

$$g_n = \sum_{\nu=1}^{n} \lambda_{\nu-1} \Delta f_\nu$$

converges a.s. in the set $\{\sup_n |\lambda_n| < \infty\}$.   p. 72

    This proof contains the elements of a proof of the martingale theorem starting from the L_2 result.

~~~~~~~~~~

The a.s. convergence of a martingale $\{f_n\}$ on the set where $\sigma(f) < \infty$ (See Remark IV 4.4 p. 160).

~~~~~~~~~~

BIBLIOGRAPHY

[1] D. G. Austin, A sample function property of martingales, Ann. Math. Stat. 37 (1966) 1396-1397.

[2] D. L. Burkholder, Distribution function inequalities for martingales, Ann. Prob. (to appear).

[3] D. L. Burkholder, Martingale transforms, Ann. Math. Stat. 37, 6, (1966) 1494-1504.

[4] D. L. Burkholder, Successive conditional expectations of an integrable function, Ann. Math. Stat. 33 (1962) 887-893.

[5] D. L. Burkholder, B. J. Davis, and R. F. Gundy, Integral inequalities for convex functions of operators on martingales, Proc. Sixth Berkeley Symp. Math. Stat. and Prob. U.C. press (1972) 223-240.

[6] D. L. Burkholder and R. F. Gundy, Boundary behaviour of harmonic functions on a half-space and Brownian motion, (to appear)

[7] D. L. Burkholder and R. F. Gundy, Distribution functions inequalities for the area integral, Studia Math. (to appear).

[8] D. L. Burkholder and R. F. Gundy, Extrapolation and interpolation of quasi linear operators on martingales, Acta. Math. 124 (1970), 249-304.

[9] D. L. Burkholder, R. F. Gundy and M. L. Silverstein, A maximal function characterization of the class H_p, A.M.S. Transactions 157, (1971) 137-153

[10] B. Davis, On the integrability of the martingale square function, Israel J. Math 8, 2 (1970) 187-190.

[11] J. L. Dobb, Stochastic Processes, J. Wiley (1953)

[12] C. Fefferman, Characterization of bounded mean oscillation, A.M.S. Bulletin 77 (1971) 587-588.

[13] C. Fefferman and E. M. Stein, H^p spaces of several variables, Acta Math. (to appear).

[14] A. M. Garsia, A convex function inequality for martingales, Ann. Prob. (to appear).

[15] A. M. Garsia, The Burgess Davis inequalities via Fefferman's inequality, Arkiv f. Math. (to appear).

[16] R. Getoor, and M. Sharpe, Conformal martingales, Inv. Math. V. 16, pp. 771-308.

BIBLIOGRAPHY

[17] R. F. Gundy, A decomposition for L^1-bounded martingales, Ann. Math. Stat. 39 (1968) 134-138.

[18] R. F. Gundy, On the class L log L, martingales and singular integrals, Studia Math. 23 (1969) 109-118.

[19] C. Herz, Bounded mean oscillation and regulated martingales, (to appear).

[20] F. John, Rotation and Strain, Comm. Pure Appl. Math. 14 (1961) 391-413.

[21] F. John and L. Nirenberg, On functions of bounded mean oscillation, Comm. Pure. Appl. Math. 14 (1961) 785-799.

[22] M. A. Krasnoselskii and Ya. B. Rutickii, Convex functions and Orlicz spaces, (Transl. L. F. Boron) P. Noordhoff LTD, Groningen (1961).

[23] P. A. Meyer, Le Dual de H^1 est BMO, (cas continu), Séminaire de Probabilités VII (U. Stras.) Springer-Verlag, Berlin (1972)

[24] P. A. Meyer, Martingales and stochastic integrals I, Springer-Verlag, Lecture notes in math series: Inst. Math. U. Stras. 284 (1972)

[25] J. Moser, On Harnack's theorem for elliptic partial differential equations. Comm. Pure Appl. Math. 14 (1961) 577-591 .

[26] J. Moser, A Harnack inequality for parabolic differential equations, Comm. Pure Appl. Math. 17 (1964) 101-134 .

[27] E. M. Stein, Note on the class L log L , Studia Math.

[28] A. M. Garsia, A presentation of Fefferman's theorem, Seminar notes.